FROST

STRESS MITIGATION IN SUBTROPICAL FRUIT ORCHARDS

NIPA GENX ELECTRONIC RESOURCES & SOLUTIONS P. LTD.
New Delhi-110 034

FROST
Stress Mitigation in Subtropical Fruit Orchards

Shashi K. Sharma, Ph. D.
Professor (Fruit Science)
Dr. Y.S. Parmar University of Horticulture and Forestry
College of Horticulture and Forestry, Neri
Hamirpur - 177 001, Himachal Pradesh, India

NIPA GENX ELECTRONIC RESOURCES & SOLUTIONS P. LTD.
New Delhi-110 034

NIPA GENX ELECTRONIC
RESOURCES & SOLUTIONS P. LTD.

101,103, Vikas Surya Plaza, CU Block
L.S.C.Market, Pitam Pura, New Delhi-110 034
Ph : +91 11 27341616, 27341717, 27341718
E-mail:newindiapublishingagency@gmail.com
www: www.nipabooks.com
For customer assistance, please contact
Phone: + 91-11-27 34 17 17 Fax: + 91-11- 27 34 16 16
E-Mail: feedbacks@nipabooks.com

ISBN: 978-93-91383-36-7

Composed and Designed by NIPA.

Dedicated
To
Sargam & Swastik

Preface

Subtropical regions are the intermediate zone between the tropical and temperate regions where winters usually experience sub-zero temperatures with intense frosty situations. The evergreen plant species suffer huge frost damage in the subtropics, especially in the low-lying basin areas and the adjoining plains. The extent of damage to the subtropical plant species largely depends on the intensity and duration of frost stress as well as the intrinsic potential of these plants to withstand the frost stress. The topographical conditions are also the major determinants of frost damage in these regions. For instance, in the NW sub-Himalayan conditions, the converging and diverging geo-morphological ranges around and adjoining the small basin areas, the surface temperature inversion is a phenomenon of common occurrence during winters. The frost events of variable intensity are quite common over here, therefore.

It was the winter of 2008-09, large areas under fruit plantations in the subtropical low hill and valley region of Himachal Pradesh suffered huge economic loss. It was realized that despite the significant advances in crop production practices, we don't have sufficient technology for fighting such meteorologically induced distress. Whatever the information or recommendations available were tried but for no use. Fruit orchards, especially those of mango, aonla, guava and litchi etc. suffered heavily. After this incidence the efforts of the government and university were accelerated for developing frost impact mitigation strategies in the subtropical fruit plantations. The Agricultural Technology Management Agencies of the frost-sensitive districts were assigned the duty of facilitating the research at the regional level. With time, few more agencies took initiatives for dealing with the frost issue. This book is, therefore, a showcase of the information generated and gathered on frost, its implications in subtropical horticulture and impact mitigation strategies.

Besides, giving an overview of frost and its impact on subtropical fruit plantations, this book is an attempt to convey the simplified version of the information on frost and related terminology. It discusses the orchard energy balance which is of utmost importance for working out strategies for energy conservation at the orchard level and effectively operating the frost protection equipments in a fruit orchard. Not only it narrates a simplified account of the mechanism of damage to plant tissues but also introduces the readers to the biomolecular aspects of plant defence against low-temperature stress. Keeping the growers perspective in mind, the practical aspects and elaborations derived from open field studies have been duly considered while describing the frost sensitivity of crops, damage symptoms, frost prediction, delineation of frost prone areas, and frost impact mitigation. For the interest of the researchers, the topic of frost tolerance development has also been touched. Active frost protection methods and their augmentation with the use of thermohysteric and hydrophobic compounds have been described in detail together with other practices. As frost is a natural hazard, despite every effort, frost damage occurs to the subtropical fruit plantations under many situations. For such situations, the growth restoration practices have also been presented in this book. This manuscript is therefore, a comprehensive narrative of theoretical and practical aspects of frost and its impact mitigation.

As we know, climate change is looming, the altered climatic patterns and increased temperature variance are expected to increase the frost risk in the subtropics, especially in the sub-Himalayan regions and the adjoining plains. The evergreen species, especially those of tropical origin, have been speculated to suffer more because of their limited adaptation to the oscillating winter temperatures. The resultant would be a reduction in the yield and profitability of the orchards. Scientists across the world are working on developing resilience to the climatic changes with enthusiasm, tenacity and dedication to keep up with this threat. The increased awareness and knowledge about these changes are considered as the key factors for crises mitigation. Good literature has always been proved instrumental in the creation of awareness and dissemination of knowledge among the masses. This book is, therefore, an initiative for developing understanding amongst the masses about frost, frost induced freezing and options which can be explored for mitigating its ill effects.

The dream of compiling this book wouldn't have been fulfilled unless the frost-related studies in subtropical fruit plantations were financed by different funding agencies. First partial funding was received from Agricultural Technology Management Agency (ATMA) Hamirpur, Himachal Pradesh. Later, the funding from the Department of Science and Technology, Govt. of India in the form of a project on 'Freeze Restriction Studies for Frost Protection in Subtropical Fruit Crops' helped continue most studies. Continuous contingent financial support by Dr. Y. S. Parmar University of Horticulture and Forestry, Solan, Himachal Pradesh served as the major salvager. Without financial assistance from these organizations, it would not have been possible to undertake extensive research on various aspects of frost and its impact on subtropical fruit plantations.

Writing this book was harder than I thought. This would have not been possible without continuous persuasion by my wife – Shailli. From providing a contemplative atmosphere at home to reading the early drafts of the book, she has contributed a lot to bringing this book to you. I owe special debt and gratitude to my parents for supporting my work.

January 2022 **Author**

Contents

List of Colour Plates

List of Colour Plates

1

Frost and Subtropical Fruit Growing – An Overview

Fruit and nut crops have adapted to a wide range of soil, climate and moisture conditions irrespective of their place of origin. The limits of the distribution of these crops across the globe are largely governed by climatic factors such as temperature, wind direction, solar luminance level, length of day (latitude) and atmospheric water vapour content besides other factors. These climatic factors are considered critical in governing the distribution of plant species because not only they limit the growth and development of plant species but also determine the consistency of annual cropping and the quality output. Therefore, the more we learn about the climate of a specific area, the better understanding of the physiological responses of the plant to that environment can be developed. The plant environment is usually referred to as factors like soil, water, light and temperature etc. concerning growth and development. Amongst these factors, soil and water can be modified to a great extent through various soil amendment and water management techniques. But, light and temperature are the factors that are difficult to control under normal and natural field conditions. Though, the light is more a latitudinal factor that exerts a photoperiodic effect at the orchard level, we can manage its photosynthetically active part through canopy architecture manipulations. The fourth environmental factor *i.e.* temperature is the most critical one, the cardinal limits of which are specific for different plants and beyond human control under the open environmental conditions. It is also critical because it regulates almost all processes of plant growth and development. The plants either malfunction or die if the temperature goes above or below the functional/cardinal limits required for it. Within the cardinal limits also, there are many physiological processes the output of which is very much affected by the temperature. For example, autumn pre-conditioning and acclimation, breaking of dormancy, and phenology etc. are the processes that get influenced greatly by the minor temperature fluctuations. Many physical processes are also influenced by temperature such as diffusion of the gases and liquids in the plants, the solubility of the ions, transpirational pull (which affects the rate of transport and transpiration). All biochemical reactions and responses in a plant system are further temperature-dependent; the rate of reactions increases as the temperature rises, but such rate increases vary with the specific reaction type.

In a natural habitat, plants are rarely injured by cold or hot weather conditions as they have evolved adaptive physiological, morphological, biochemical or anatomical mechanisms that permit them to survive the climatic extremities. For example, the leaves of high-latitude (boreal) species sense the shortening of day length in late summer and initiate inhibitory mechanisms that cause the plants to stop growth well in advance of the first heavy frost of winter. Another mechanism in the plant system induces cold acclimation in response to the initial low-intensity frosts. Mid-latitude temperate plant species have yet developed a third kind of adaptive physiology, *i.e.* incorporating some of the characteristics of both high and low latitude species. The winters in the middle latitudes may fluctuate between cold and mild temperatures. The temperate species thus have developed long rest-period chilling requirements; they will not resume growth even when warm growing temperatures prevail for a few days during their winter rest. On the other hand, in tropical climates, there is normally no freezing as the temperature rarely falls below 10^{0}C and sometimes if, temperature falls below 7^{0}C chilling injury may be observed in the evergreen species in the tropics. But, at the subtropical latitudes (23.5^0 to 35^0) the night temperature during winter may fall below the freezing point, thus the frosts of variable intensities are quite common in these regions. The frost causes freezing and induces injury to the plants which tend to continue their growth because of the favourable daytime temperature and the soil moisture. Usually, the evergreen plant species in these regions do not get induced to the proper acclimation process, a defence mechanism against frost and low-temperature stress. The accumulation of cool air at certain pockets further aggravates the freezing environment and leads to huge damage to the subtropical evergreen species under such agro-ecological conditions.

The extent of damage to the plants due to frost-induced freezing depends mainly upon the intensity and duration of frost-induced freezing stress, which apart from the latitudinal factors depends upon the local topographical conditions, also. For instance, the low hill and valley region in the Northern India comprises several ranges roughly going parallel while converging and diverging at places giving rise to undulating topography of hills/hillocks surrounding small basin area in between. These ranges arise gradually from adjoining plains and go on attaining a height of 300 to 1100 m above mean sea level. Owing to these topographical conditions, the surface temperature inversion is a phenomenon of common occurrence during winters in these areas. The region is highly risky for growing frost-sensitive fruit plant species; despite the subtropical latitudes, it is sub-optimal for growing of subtropical fruits. The average minimum temperature of the region remains below 7°C for about four months during winters. The temperature below freezing is also not uncommon during winters in the region. Low-lying plain or basin areas experience pooling of cold air due to continuous accumulation of cool air near the earth surface, that makes it vulnerable to frost and low-temperature stress.

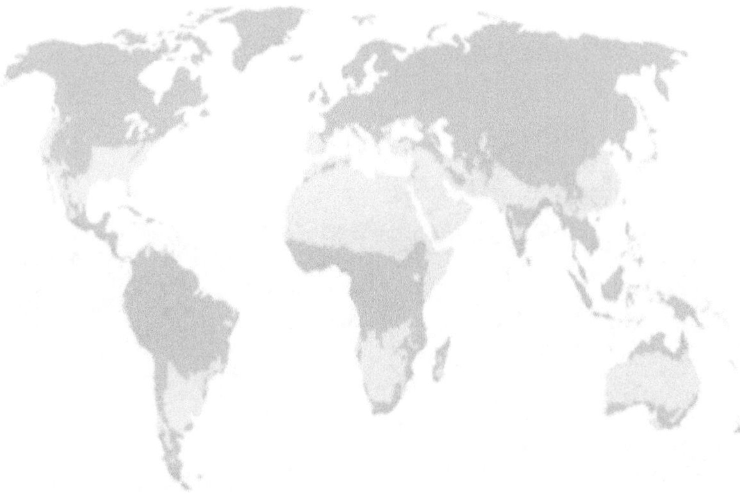

Pic. 1: Areas World with Subtropical Climate
Source: Koppen [1]

Most of the devastating frost events are experienced mostly during the month of January in the subtropical regions of NW India; sometimes in December and February too. The early and late frosts are also dangerous for subtropical plants as these inflict heavy damage to the foliage, twigs, branches and flower buds etc. The evergreen subtropical fruit species like papaya, banana, mango, litchi, aonla, guava etc. suffer huge economic loss every year in the low hill and valley region of NW India.

As fruit growing is a capital-intensive and long-term enterprise, the establishment of a fruit orchard, especially of subtropical fruit species is always challenging under frost-prone situations. It incurs heavy losses to the growers years after year. Fruit growing is a compulsive agricultural avenue in the subtropics as these areas are densely populated and there is high pressure on the landholdings for drawing maximum output for sustaining the livelihood. In the past decade, there has been observed an increase of about 172% in the area under subtropical fruits in Himachal Pradesh; but, the corresponding increase in production and productivity of these crops is still very poor as these crops suffer more economic losses due to frost than caused by any other weather hazard. In a recent estimate, it has been reported that during a frosty year there occur direct losses of about one hundred and thirteen million rupees per annum in the NW subtropical region of India [2]. These are only the observable damage losses (exact loss estimates are difficult to make), the impact of frost on the farmers and local economy is so devastating that people are cutting an edge from horticulture and associated activities due to this natural hazard.

Pic. 2: Frost Affected Mango Orchard

Even though it is a fact that frost is devastating and its occurrence is common in NW India, sufficient technology for its impact mitigation is still lacking. We made impressive advances in production technology over the last few decades, but fruit production in the subtropical regions is still governed by the observance of a frost year or a non-frost year. The results of climate change studies across the globe have shown that the increase in the mean temperature has reduced the risk of frost damage, as expected; but, the increase in the variance of temperature increases this risk. Specifically, an increase in temperature of 1°C increases the chances of successful fruiting of plants by five per cent; but, the corresponding temperature variance increase by 10% decreases the chances of good fruit production by the same percentage. As per Rigby and Porporato [3], "If the basic degree-day models are correct, temperature variance has as much of an effect on plants as changes in the mean". Therefore, if the temperature becomes more erratic, the temperature on any given day is less predictable from the temperature of the previous day, this increases the risk of frost damage to plants. Thus, it is a reality that climatic variability is going to play a greater role in the years to come. Therefore, the vulnerability of the subtropical regions to frost is expected to increase in the near future. Development of comprehensive frost impact mitigation strategy is therefore urgently needed. It will not only benefit the orchardists but also influence the economy at the local as well as at the regional or national level. Every individual who has an association with subtropical fruits, not only at the production level but also at trading or consumption levels, can be benefited through the technological advancement in the handling of frost and its devastating effects. And, for technological empowerment; the development of honest literature is very essential, so are the efforts of this book.

2

Frost Definition, Types and the Related Terminology

Frost is a thermodynamic phenomenon that occurs in the atmosphere, especially under subtropical and temperate conditions during winter. To understand the horticultural implications of this meteorological event, it is necessary to understand the frost related terms and jargon, first. Some of the common terms used here in this book are described hereunder:

Frost: Definitions and Types

'Frost' is generally perceived as the solid deposition of water vapours from saturated air; it is formed normally when a solid surface in contact with air is cooled below the deposition point (frost point). Specific heat, the thermal emissivity of the surfaces and the number of water vapours present in the air define the intensity of the frost. It is also affected by the difference in thermal absorptivity of the surface and its specific heat which under non-windy situations greatly influence the temperature attained by the super-incumbent air. As the cold air is denser than warm air; under calm weather, it pools at the ground level. This is known as **'surface temperature inversion'**; and this is the reason why frost is more common and extensive in low-lying areas. Frost can form in these areas even when the reported air temperature is above the freezing point of water.

Hogg [4, 5] and Lawrence [6] defined frost as an occurrence of temperature below 0°C measured in a 'Stevenson-Screen' shelter at a height between 1.25 and 2.0 m above ground. Cunha [7] stated that frost is a vapour freezing condition when the surface temperature falls below 0°C. From a crop production point of view, the frost may be defined as a meteorological event when crops of the subtropical region suffer freezing injury. The word frost and freeze are generally used interchangeably to describe the drop in air temperature to near 0°C or below when the formation of ice crystals on the surfaces takes place either by freezing of dew or phase change of water vapours to ice.

As the perception of frost was different to different researchers, they defined it differently, therefore. For FAO, Richard *et al.*[8] presented a generalized definition of frost as under:

> *A 'Frost' is the occurrence of an air temperature of 0^0C or lower (measured at a height of between 1.25 and 2.0m above ground level, inside an appropriate weather shelter). Water in the plant may or may not freeze during a frost event, depending upon several avoidance factors (e.g. super-cooling and the presence of ice-nucleating factor). A 'freeze' occurs when extracellular water in the plant freezes. This may or may not lead to damage of plant tissue, depending upon tolerance factors (e.g. solute content of the cells). A frost event becomes a freeze event when extracellular ice forms inside the plant tissues. Freeze injury occurs when the plant tissue temperature falls below a critical value where there is an irreversible physiological condition that is conducive to the death or malfunction of the plant cells. This damaging plant tissue temperature is correlated with air temperature called 'critical temperature' measured in standard instrument shelters. Subzero air temperatures are caused by a reduction in the sensible heat content of the air near the surface, mainly resulting from a net energy loss through radiation from the surface to the sky and/or wind blowing in subzero air to replace warmer air.*

More conveniently, in meteorology, frost is the formation of ice crystals on surfaces due to deposition, that is a phase change from vapour to ice. In agriculture or biology, however, frost is an event where the temperature falls to the point where ice forms inside plant tissues and causes damage to the cells.

Frost Types

As stated above, the frosty conditions generally appear either due to rapid irradiance of the exposed surfaces during a cool night or due to accumulation of cool air in low-lying areas leading to temperature inversion. The frost is thus categorized into two main categories viz. radiation frost and advection frost.

1. Radiation frost: Radiation frost is a characteristic feature of nights that have a clear sky, clam with low or no wind velocity, low dew point temperature and air temperature near 0^0C or below. The daytime temperature during such events is quite high, there exists a wide difference in the day and night temperatures. There occurs a sharp drop in temperature at a negative exponential rate after the sunset.

During a radiation frost event, the heat is lost into the open sky causing objects to become colder than the surrounding air and the water vapours present in the air freeze onto the rapidly cooling surface in the form of white ice crystals, loosely deposited on the ground or exposed objects. It is also known as **'Hoar Frost'** or **'White Frost'.** As the temperature falls faster near the radiating surface, it causes a temperature inversion (i.e. the temperature increases as we move above the ground level); it has been observed that the normal surface temperature of soil (depending upon the emissivity of the surface) always remain 2 to 3^{0}C lower to the air temperature measured in a meteorological equipment shelter at a height between 1.25 and 2.0 m above ground. This occurs due to net loss of energy through radiations from the surface, the sensible heat content of the soil surface and the air near the surface decreases. A flux of sensible heat gets generated downward from air to soil surface and upward from within the soil to the surface to replace the lost sensible heat. This result into temperature decrease aloft as well, but not as rapidly as at the surface. The air depth to which the temperature inversion is effective is variable and depends upon the local topography and weather condition. In the low hill and valley region of Himachal Pradesh, it has been observed that the major temperature inversion occurs up to about a vertical height of 100 meters from the lowest point of a micro-watershed [154]. This type of conditions lead to the formation of frost which is called **'Ground Frost'** or **'Grass Frost';** when seen on the ground or grass, respectively.

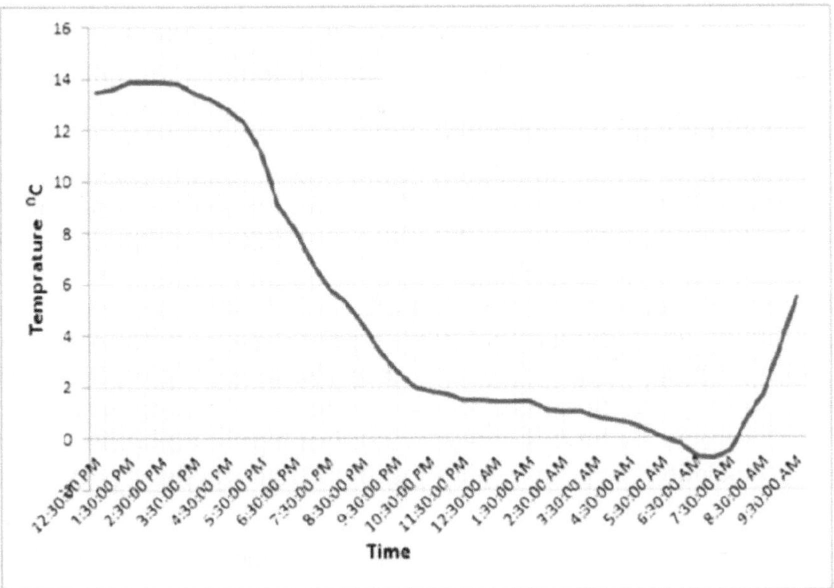

Fig. 1: Temperature drop during a typical radiation frost event

Apart from the hoar frost, there occurs another type of radiation frost called **'Black Frost'.** It is such a low temperature condition when the air temperature falls below 0^0C but ice formation does not take place due to low atmospheric humidity. As ice formation is an exothermic process, some amount of heat (80 calories per gram of water frozen) is dissipated into the atmosphere in case of hoar frost, but there is no such exchange of heat under black frost conditions, therefore black frost causes more damage to the plants than hoar frost. There are a few other terms that are used worldwide to describe practically experienced radiation frost conditions. For example, in the cold regions, there has been experienced condensation and subsequent freezing of water vapours onto a window glass surface which is exposed to extreme cold weather outside and comparatively a moist & warmer atmosphere inside. Popularly such a type of frost is called **'Window Frost'**. In many cases, such type of incidence is more a condensation phenomenon and rarely happens to be a frost event, except under extremely cold conditions. Under very cold, wet and windy weather there has been observed quick solid ice crystal formation which is known as **'Rime or Glaze'**. The rime is usually seen on the exposed surface of the ships travelling through the Arctic Ocean. Technically speaking, it is not a type of frost since usually super-cooled water drops are involved, in contrast to the formation of hoar frost, in which water vapour de-sublimates slowly and directly.

2. Advection/ Convection/ Pool Frost: Advection Frost is defined as the frost that usually occurs due to the horizontal transport of cool air from one place to another by bulk fluidic motion. For example, when the cool air from hills blows into adjoining areas, it drops the temperature of these areas leading to the formation of frost. On the other hand, **convection** is a process of movement of cool air in response to heat. The most common example of convection frost is the cooling of the air near the Earth surface and draining of this cool air down the slope in hilly areas leading to accumulation of cool air in the low-lying area leading to the occurrence of frost in these areas. The movement of cool air toward the Earth surface and subsequent lifting of warmer air to the upper horizon is also a type of convection. The occurrence of frost with such type of heat transfer mechanisms is called **convection frost**. Most of the times, the terms advection and convection are used interchangeably. Both types of these atmospheric phenomena when lead to the accumulation and pooling of cool air in a certain low-lying pocket area, it is called **pool frost**.

As said above, advection convection frost involve the movement of cold air that blows into an area to replace warmer air, it may or may not be associated with cloudy conditions, moderate to strong winds, no temperature inversion and low humidity. Often temperature drops to freezing point and stays there for a longer time (for a day or more). Usually, a continuum of nights with subzero temperature starts as an advection frost which later gets changed to radiation frost nights. Under pool frost conditions the wind speeds may be very low, the sky may be cloudy, but the temperature still falls to a minimum well below 0°C. The impact of such a frost reduces as we move up the slope in the hilly terrain. In low-lying plain areas, if it persists for three or four days in continuation, it is likely to persist further for three to four more days despite the sunny weather at the higher reaches; as by the time the thick blanket of cool air gets heated up, the sunset arrives and again the cooling process begins. Therefore, this type of frost persists for a longer duration and the continuous persistence of foggy and frosty conditions lead to heavy loss to the subtropical crops due to impaired plant physiological functions and hypothermia.

Sensible Heat

When an object is heated, its temperature rises as the heat is added, this increase in heat (temperature) of the object is called sensible heat. In other words, the heat which causes a temperature change is sensible heat. In more simple terms, the heat which can be sensed is called sensible heat and therefore, the heat which we measure with the thermometer is called sensible heat.

Latent Heat

It is the chemical energy that is stored in a compound in the forms of chemical bonds holding its molecules together. In water, a lot of energy is stored in the forms of bonds holding the water molecules together. Upon phase change, when water condenses, cools, or freezes, the rearrangement of the water molecules dissipates their chemical energy to the surrounding in the form of heat; this energy released is called as the latent heat of the water. This change of latent heat to sensible heat occurs mainly upon phase change of water from liquid to solid or vapour state or vice versa. When water changes its phase from liquid to solid (ice) or vapour, the latent heat gets changed to sensible heat and the temperature of the air in the surrounding rises. When the reverse happens, *i.e.* when the ice melts, the water warms, or water evaporates, sensible heat is changed to latent heat and the air temperature in the surroundings falls.

Around 334 Joule (79.9 calories per gram) of energy are required to melt 1g of ice at 0°C, it is called the latent heat of melting; it implies that at 0°C liquid water has 334 Joule of energy more than ice at the same temperature, this is called **Latent Heat of Melting or Latent Heat of Fusion.** This energy is released when the liquid water freezes at 0°C. The amount of heat which is required for the vaporization of water at 100°C is about 2,230 Joules (533 calories) per gram and we know it as **Latent Heat of Vaporization**. When water is kept boiling in a pot, its temperature remains at 100°C until the last drop evaporates because all the heat being added to the liquid is absorbed as latent heat of vaporization and carried away by the escaping vapour molecules. Similarly, while the ice melts, it remains at 0°C, and the liquid water that is formed with the latent heat of fusion is also at 0°C and 80 calories per gram get added to the surroundings of the system.

Temperature Inversion

During irradiative cooling of the earth surface through the night, there occurs a net loss of energy in the form of long wave radiation from the surface to the sky which acts as a black hole and absorbs all the radiations coming toward it. During this cooling event, the sensible heat content of the air decreases near the earth surface. Due to a deficit of sensible heat near the surface, heat transfer occurs downward (by convection) from the air above to partially replace the surface heat loss. This transfer of sensible heat from the upper layers of air to the lower ones is called temperature inversion. As the irradiative cooling continues, the temperature inversion also continues for the whole night. When the Sun rises in the morning, the surface begins to get heated again and the inversion gets stopped.

Super-saturation

It is a state when a solution contains more of the dissolved material than what could be dissolved by the solvent under normal circumstances. In the case of atmospheric humidity, super-saturation refers to the presence of excess water vapours in the air that it can hold at a given temperature and pressure. Super-saturation is thermodynamically an unstable state which gets relieved as the condensation or precipitation takes place.

Dew Point

It is the temperature at which the surface cools enough to cause precipitation or condensation of the water vapours present in the saturated air. At the dew point, the number of water molecules striking the surface and forming hydrogen bonds with other water molecules is bigger than the number of molecules breaking hydrogen bonds and separating off as a gas. Technically, the dew point is defined as the temperature when air is cooled adiabatically (with no outside sources or losses of energy) until it becomes saturated enough to relieve water vapours through condensation.

Frost Point

Frost point is similar to the dew point; the difference is, at the frost point the phase of water present as water vapour in the air directly changes to the solid phase (ice) without going into a liquid phase. Hence, the frost point is the temperature at which the water vapours present in the saturated air freeze directly onto a cooling surface without undergoing condensation. For water molecules, it is more difficult to escape a frozen surface than a liquid surface as, ice has a stronger affinity for neighbouring water molecules than water. This is why the frost point is greater in temperature than the dew point; or, why sometimes frost occurs above 0°C air temperature.

Super-cooling

It is the cooling of liquid or water below the freezing point without change of phase to solid (ice). More conveniently, it can be said that the maintaining liquid state of water below the freezing point is called super-cooling. A liquid below its freezing point generally crystallizes in the presence of a seed crystal or nucleus around which a crystal structure can form. Under super-cooling, the formation of the nucleus crystal is restricted and the liquid phase can be maintained down to the temperature at which homogeneous ice nucleation occurs.

Ice Nucleation

The melting point of water (transformation from solid to liquid phase) is 0°C, but it does not necessarily means that it also freezes (transformation from liquid to solid phase) at 0°C. For freezing to occur, ice nucleation either homogeneous or heterogeneous is essentially required. As the ice nucleus is formed, it attracts the water molecules from the surrounding, leading to the growth of the ice crystal. The process of ice nucleus formation is called ice nucleation. In the case of pure water, ice nucleation takes place without the influence of other material forces at -40°C and this type of ice nucleation is called homogenous ice nucleation. At temperatures above −40°C (−40°F), the ice nucleation can take place with the presence of other materials dissolved in water or due to the presence of other physical, chemical or biological agents which facilitate ice nucleation even at 0°C. Such a type of ice nucleation is called heterogeneous ice nucleation. In nature, the heterogeneous nucleation is facilitated by the presence of silicate minerals of terrestrial origin, mainly clay minerals and micas. Organic compounds like humus and ice-nucleating bacteria also play an important role in heterogeneous ice nucleation. Heterogeneous ice nucleation is more common, but homogenous mechanisms become more likely as the degree of super-saturation or super-cooling increases.

The frost-induced freezing in plants gets initiated in two ways; either ice nucleation occurs inside the plant system or it occurs outside, at the surface of the plant. Under the conditions of low atmospheric humidity when the temperature goes below the freezing point in many plants, the ice nucleation gets initiated first inside the plant tissues, this type of ice nucleation is called **intrinsic ice nucleation**. However, in highly frost-sensitive plant species like papaya, banana etc.; intrinsic ice nucleation may take place irrespective of the atmospheric humidity. Otherwise, under most of the conditions when atmospheric humidity is normal or above, as the surface temperature approaches freezing, the water from saturated air forms the ice nucleus at the cooling surface and this type of ice nucleation is called **extrinsic ice nucleation**. The externally formed ice then makes its entry into the plant system through stomata, hydathodes or other openings or wounds and cause freezing of intracellular water and then progress further to other cellular compartments.

Thermal Hysteresis

This term is generally used in relation to the difference between the freezing temperature and the melting temperature of a substance or compound. As the case of pure water is; it melts at 0^0C, but super-cools to about -40^0C in bulk or even lower if no ice nucleators are present on the cooling surface. This property of variable freezing or melting behaviour of compounds is known as thermal hysteresis. More simply, it can be said that the compounds which induce freezing, melting or boiling point alteration/ depression are called thermo-hysteric compounds. Glycols or alcohols are the most common thermal hysteresis inducing compounds that are used in coolants, ice creams or other food products for super-cooling and prevention of ice crystal formation or for changing the boiling and freezing point of the fluids.

Low Temperature Damage

Temperature is the primary factor affecting the rate of plant development and limits the productivity and distribution of plants across the globe. Both high and low temperatures beyond the cardinal limits of the plant induce a physiological disturbance in the plant system, which is usually called stress. And, when stress leads to physical disruption of plant structural mechanism, it is called injury or damage. Among various environmental stresses, low-temperature stress is one of the most critical stresses as it has multiple implications on the physical and physiological aspects of the plants. When the temperature drops below the normal limit, it induces stress which may cause injury to the plants. Low-temperature stress usually results in two types of injuries in plants: i) Chilling injury and ii) Frost or Freezing injury.

i) Chilling Injury: It is the damage or injury caused to the plant or plant tissues by low-temperature, above 0°C. In tropical plant species, a chilling injury may be caused by temperature ranging from 0-13°C; for example, Banana fruits get damaged at temperatures below 13°C. In the case of subtropical plants, chilling injury usually occurs at temperatures 0-7°C; therefore, the temperature of 0 to 7°C is considered as a chilling temperature and the injury which is caused due to exposure to this temperature range is known as chilling injury.

Chilling sensitive tropical and subtropical plants exhibit a marked physiological dysfunction when exposed to non-freezing low temperatures; it may lead to the death of a plant or plant part depending upon the temperature and duration of exposure to the chilling temperature. Low temperature causes disturbances in all physiological processes – water regime, mineral nutrition, photosynthesis, respiration and metabolism etc., all get affected. It is the change in the rate of gene transcription of low molecular weight proteins which determines the response of the plants to low temperature exposure.

The symptoms of chilling injury vary with the plant type, tissue and severity of the injury. The symptoms usually develop more rapidly if the injured tissue is transferred to a higher non-chilling temperature after low temperature exposure. Some of the most apparent symptoms are surface pitting, necrotic areas, external discolouration, brown spotting etc., these may vary with the age of the tissue, also. For examples, when green bananas are exposed to mild chilling temperature; the peel develops a smoky or dull-yellow appearance whereas, when exposed to severe chilling temperature, it turns black or brown. Surface pitting is common in cucumber. The internal and external discolouration is common in sweet potato, when it is cut and exposed to a short period of chilling. In many of the subtropical and tropical fruits, the mature fruits fail to ripen when exposed to chilling temperature, e.g., under the Himalayan subtropical conditions the winter crop of mature green guava fruits fails to ripen after mid of November if the daily minimum temperature touches the chilling range, the typical climacteric rise in respiration is not observed in these fruits. Disturbances in ethylene evolution have also been reported for chilled bananas, grapefruit, sweet oranges, avocados in storage. Wooliness of peaches and plums refers to a mealy texture and discoloured appearance of the flesh caused by chilling injury in storage at 0° to 4°C. The tissues damaged due to chilling rapidly get invaded by the decaying microorganism and therefore, the chilling injury is many times confused with microbial infections also.

ii) Freezing Injury: It is the injury caused to the plant or plant tissue by a temperature of 0°C or below. Both chilling and freezing injuries are related to low temperature damage, the difference is that freezing injury is caused at freezing temperature or below and associated with inter or intracellular ice crystal formation in the plant tissues. In the case of chilling injury no ice crystal formation occurs, the damage to the tissues is more due to cellular dysfunctions rather than cellular disintegration and disruption caused due to ice crystal formation which is common in the case of freezing injury.

Cellular dehydration is one of the most common effects of freezing injury caused due to extraction of cellular water by the growing ice crystal in between the intercellular spaces. As the growing ice molecule possesses high affinity of water molecules, it attracts the surrounding water molecules to grow itself. This type of damage is sometimes reverted through the proper thawing process. Another effect of freezing is cellular disintegration or disruption which is caused due to growth of ice crystal within the cell which leads to the breakage of cellular membranes. It may lead to the complete disintegration of the cellular contents. The damage so caused is irreversible. The tissues injured due to such type of a freezing injury, upon thawing, results in symptoms like that of water-soaked areas, mushy texture and complete collapse/breakdown of the damaged areas.

Frost Injury

Frost formation and plant tissue damage are not related. But, during a frost event when the ice crystals formed on the surface of the plants, they make their way into the plant system causing either intercellular or intracellular ice crystal formation leading to several types of damages like cellular dehydration (due to intercellular ice formation) or cellular disintegration or disruption (usually due to intracellular ice crystal formation). Frost thus induces freeze injury and therefore, the terms frost injury and freeze injury are generally used interchangeably in the context of frost related studies. One notable thing is that the frost may occur at air temperature above 0°C. The occurrence of frost is related more to temperature of the cooling surface and the dew point/frost point rather than the temperature of the air. But, frost injury usually occurs when the surface temperature dips below freezing even for a short period. The severity of the damage depends on how long and how low the surface temperature actually goes below freezing. Frost injury is discussed in detail in the forthcoming chapters of this book.

3

Frost and Orchard Energy Balance

Earlier, it was a general perception that frost is something like snow that falls from the sky onto the Earth and causes cooling of the plant surface leading to freezing and lethal damage. But, with advancement in knowledge, it is now well understood that it is the cooling of a surface to a subzero temperature which leads to solidification of the water vapours present in the air onto it resulting in the formation of a white crystalline layer of ice molecules. The ice so formed further grows into the plant system causing damage to it through intracellular or intercellular freezing.

As frost is a thermodynamic atmospheric phenomenon, to understand the process of frost formation and subsequent frost induced freeze damage to the fruit plantations, it is essential to know about the mechanism of heat transfer within the orchard ecosystem. The intensity of radiation or pool frost within an orchard system largely depends upon the rate of heat transfer within the orchard. Under the natural conditions there are observed four main **heat transfer modes** during a frost event [8]; radiation, conduction (or soil heat flux), advection or convection (movement of cool air from one place to another) and the phase change associated with water (latent heat). The **radiation** mode of heat transfer includes the sum total of energy that is received from the Sun during the daytime. It is the only source that imparts the highest energy to an orchard system; there occurs no significant gain in orchard energy after sunset. On a cool, clear and calm night, there occurs a loss of radiations from the Earth and its occupants in the form of infrared long wave radiations. *For the frost to occur, the loss of radiation energy during the night must exceed the gain of radiation energy during the daytime.* Thus, the **Net Radiation (E_{Rn})** is an important factor for the frost to occur. And, the components which determine E_{Rn} are the E_{Rsd} (short wave solar radiations-downward – a major amount of solar energy received by earth surface), E_{Rld} (long wave radiations downward-solar radiations which reach earth surface after losing considerable energy in the atmosphere), E_{Rsu} (short wave radiations radiated upward - a little amount of radiations reflected back) and E_{Rlu} (long wave radiations upward - irradiance from Earth surface).

Conduction is another heat transfer mode in an orchard ecosystem. Some amount of energy is either added or removed from the orchard system from the lower layers of the soil to the soil surface or vice versa. This energy is conducted from molecule to molecule in the soil layers. This type of energy transfer is called **Soil Heat Flux Density** ***(G_{fd})*** at the soil surface. Apart from radiations and conduction, there is a considerable transfer of heat from one place to another by air through the processes of advection or convection. The feeling of cold which we sense is due to the striking of cool air molecules with our body. The higher the movement of the cool air molecules, the higher cold feeling we perceive. This transfer of sensible heat from one place to another through the air is called **Sensible Heat Flux Density** ***(E_H)***.

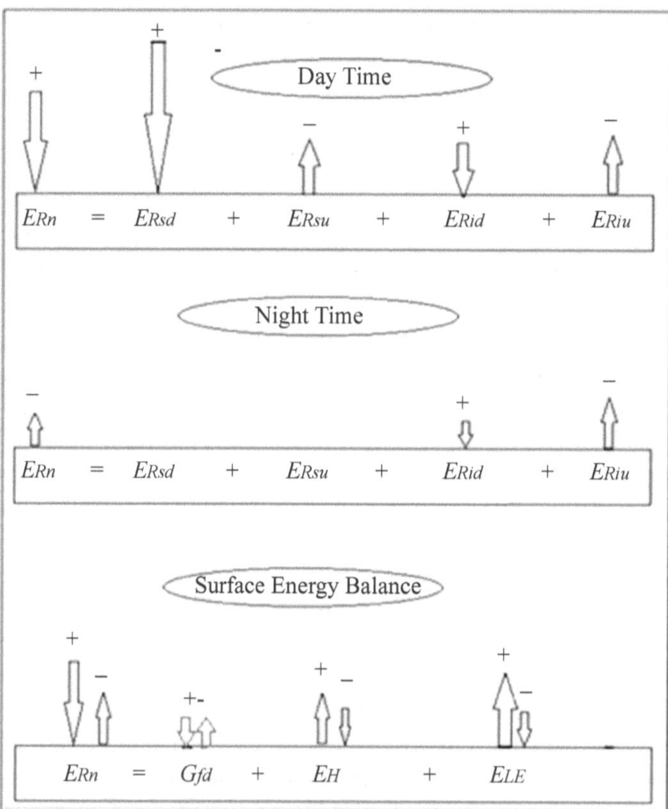

Fig. 2: Energy balance during a radiation frost event
Source: Snyder and de Melo-Abreu[8]

Apart from the sky, soil and the air; there is another component of the ecosystem that contributes to the energy balance of an orchard. This component is water, may be present as such on the surface or present in the air as water vapours.

Water possesses a good amount of energy as latent heat which is the potential energy present in water. This energy gets released into the air upon freezing of the water present as water vapours, it implies that the latent heat gets transferred to sensible heat when phase change of water from vapour to liquid or water to ice occurs. This energy transformation is known as **Water Vapour Flux Density** (**E_{LE}**) and that depends on the Latent Heat of Vaporization of water. The water vapour content of the air therefore is a measure of latent heat content. So, the humidity and its energy relation are important factors to be taken into consideration while studying the energy balance of a particular ecosystem. The energy balance of an orchard system during a radiation frost event is depicted pictorially in Figure 2. The balance energy and heat transfer between different components have been depicted with relevant notations, positive and negative signs have been used to indicate the direction of energy flow to or from the surface. Any radiation coming to the earth surface and adding to surface energy has been depicted with '+' sign whereas the radiations moving away from the earth surface and reducing surface energy has been depicted with '-' sign. For example, downward shortwave radiations from Sun and sky (E_{Rsd}) are positive, whereas short wave radiations that are reflected upward from the surface (E_{Rsu}) are negative. The 'Net Radiation' (E_{Rn}) is the 'net' amount of radiant energy that is retained by the surface after all the gain and loss of radiations to and from the earth surface. The day and night time energy balance depicted in this figure shows that if the sum total of the different components of energy happens to be positive, as it happens during daytime, then E_{Rn} is positive and more energy from radiations is gained than lost from the surface. And, if the sum total of the component is negative, as it happens during the night, then the E_{Rn} will be negative and more radiation energy is lost than gained. The energy which is used for heating of air, plants or evaporation of water is supplied by E_{Rn}. The energy balance depicted in the figure shows that energy from E_{Rn} is portioned into components G_{fd}, E_H and E_{LE}, therefore E_{Rn} is set equal to the sum total of G_{fd}, E_H and E_{LE}. Radiation adds energy to the surface, so it is depicted positively. When G_{fd} is positive, the energy is going into the soil, and when E_H and E_{LE} are positive, the energy movement is upward to the atmosphere. Therefore G_{fd}, E_H and E_{LE} fluxes are positive away from the surface and negative toward the surface. During a cool, calm and clear night, the sky acts like a black hole and keeps on sucking all the energy from the earth's surface thereby the negative components weigh out the positive components and the energy loss from the surface exceeds the energy gain resulting in the radiative cooling of the surface leading to the evolution of a frost events.

As a preventive measure of frost induced freeze damage, we can either slow down the rate of energy transfer during the night or enhance the amount E_{Rsd} by harnessing more and more solar energy during daytime at the soil surface so that the E_{Rn} could be managed positive for a longer period during night time. As the level of frost induced freeze damage to the plants is largely a function of the duration of low temperature exposure, therefore it can be said that the level of damage depends upon the fact that how fast the E_{Rn} is converted from positive to negative during the night. It implies that extension of this conversion time can avoid or reduce the frost induced freeze damage to the many plant species in a significant manner. Therefore, any factor or practice which can maintain the positivity of E_{Rn} can help in the protection of the plants from frost damage. For example, under non-windy conditions, the addition of water to the orchard system increases the E_{LE} component which favours the positivity of E_{Rn} and therefore contributes greatly for the prevention of frost induced freeze damage to the plants especially under cool and calm situations. Contrarily, under windy situations the vaporization of water being an endothermic process uses the energy equivalent to E_{LE} and the value of E_{Rn} decreases rapidly. Therefore, the addition of water to an orchard system under windy situations may cause more damage to the plants than giving protection against frost. There are many practices or methods by which the orchardists can maintain positive level of E_{Rn} in their orchards. Additions of heat through heaters, through inflow of warmer air with the help of windmills are the effective methods for maintaining positive energy balance of an orchard. Further, the positivity of E_{Rn} can also be sustained longer by improving the heat transfer and storage in the soil during the daytime and reducing the loss of G_{fd} at night; this helps for reducing the magnitude of negative E_{Rn}. The cooling or freezing water which converts latent heat to sensible heat also raises the temperature of the surrounding air, it can also help to maintain the positivity of the orchard energy. The addition of heat through radiations or convection also helps in faster reclamation of positivity of E_{Rn} because the rate of the temperature drop decreases as the temperature is raised in the orchard. Thus, a proper understanding of the energy flow and its balance within an orchard are critical from frost protection point of view. Orchardists can employ a method of frost protection effectively only if they have a sound understanding of orchard energy components.

4

Mechanism of Frost Damage

As discussed in previous chapters, the damage to the plants due to frost occurs as a result of frost-induced freezing and not merely because of the low temperature exposure. The cells of sensitive plant species get injured when the temperature drops below a critical level. When the damage caused is associated with sub-zero temperature, it is called freezing injury. And, when it is caused due to a temperature above 0^0C, it is called chilling injury. Chilling injury results in the malfunctioning of physiological processes of the plant without ice crystal formation inside or outside the plant cells, whereas freezing injury is associated with inter or intracellular ice crystal formation and usually associated with irreversible damage to the plant tissues. Plants of tropical and subtropical origin are highly susceptible to frost induced freeze injury. Frost may appear on the surface of these plants as ice crystals that may form at temperatures above or equivalent to freezing temperature (under many situations it may be a temperature equivalent of dew forming). Simple formation of frost at the surface does not damage plants. It is the growth of ice crystals from plant surface to the plant tissues which cause damage. Formation of frost takes place at the plant surface (extrinsic ice nucleation), the newly formed ice crystal need entry into the plant system to cause damage. This entry of the ice crystal into the plant system is provided by the transcuticular pores (i.e. pores between cell structures), stomata, hydathodes and/or wounds etc. This mechanism of induction and initiation of freezing inside the plant parts is quite common in several subtropical fruit species under frosty conditions. In the case of highly frost-sensitive plant species or under the situations of low atmospheric humidity, the intrinsic ice nucleation may take place in the intercellular spaces with or without external ice crystal formation. In this case entry of external ice crystals into the plant system is not essentially required for internal freezing. In both, intrinsic and extrinsic ice nucleation; the ice crystal formed, whether in the intercellular spaces or within the cell, grows in size by attracting the water molecules from surroundings. From here the process of damage to the plant system initiates.

The frost and low temperature freeze injury in subtropical plant species occur mainly, either through one or all of the following basic mechanisms:

i) Freezing of water present in the intercellular spaces and subsequent cellular plasmolysis

ii) Rupturing of cellular membranes due to freezing of intracallular water

iii) Denaturation of the sensitive proteins

iv) Dehydration of plants due to freezing of soil water

Under freezing stress; as the ice nucleation occurs, the ice crystal starts to grow in the intercellular spaces (**Intercellular Freezing**) and crystallization of all intercellular water takes place. This ice formation leads to **hyper-osmotic water stress** which in turn steadily draws water from the cellular components leading to plasmolysis of the cells. Cellular plasmolysis is a typical response of plant cells exposed to hyper-osmotic stress it leads to loss of cellular turgor [9]. The loss of turgor causes violent detachment of the living protoplast from the cell wall. Though the plasmolysis is a reversible process (**de-plasmolysis**) and the original state of the cells can be restored with hydration of the cells but under severe and prolonged low-temperature stress, in many of the subtropical species it rarely happens and denaturation of the necessary proteins and enzymatic system occurs due to dehydration. Therefore, under a natural frost event it is not the cold temperature but, ice formation that causes the actual injuries to the plants. The illustration of inter and intra-cellular freezing given by Moor [10] has been presented in the Picture 3.

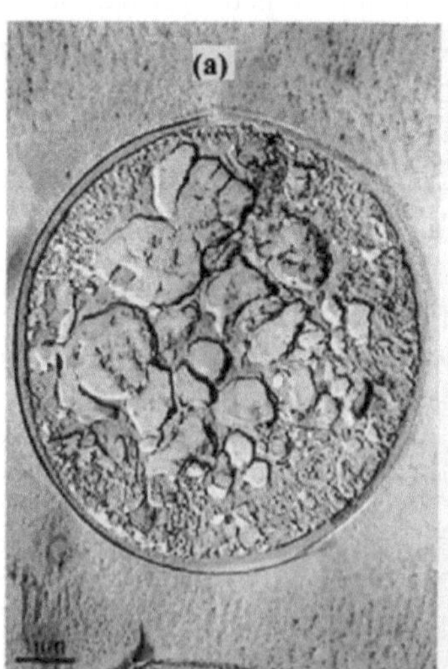

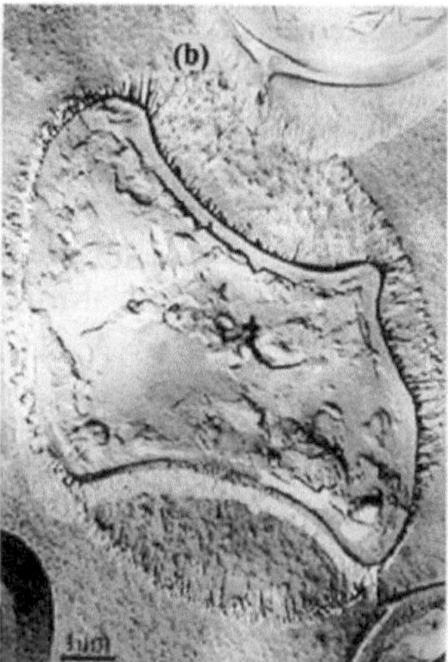

Pic. 3: a. Intracellular freezing-disruption of internal membranes and protoplasmic structure. **b.** Extracelluar freezing-shrinking of cell and its internal dehydration
Source: Moor[10]

As the surface temperature drops down to freezing level during frosty conditions, the cellular water tries to super-cool and the freezing point of cell sap is depressed by osmotically active material. The ice initially forms within intercellular spaces (apoplast) where solute concentration is the least. Water potential in that region is immediately lowered and water molecules migrate from symplast to apoplast across plasmalemma membranes and towards regions of ice crystallization, leading to intracellular ice formation. Due to the high affinity of the ice crystal for water molecules, it begins to draw the water from the surrounding cells which cause dehydration of cellular content and mechanical disruption of the protoplasmic structure[11]. As water continues to be removed from the cells, the solute concentration increases and reduces the chances of freezing. However, as ice continues to grow the cells become more desiccated. The extracellular ice crystals grow much larger than the surrounding desiccated cells. Therefore, the main cause of frost damage to plants in nature is extracellular ice crystal formation that causes secondary water stress to the surrounding cells[12].

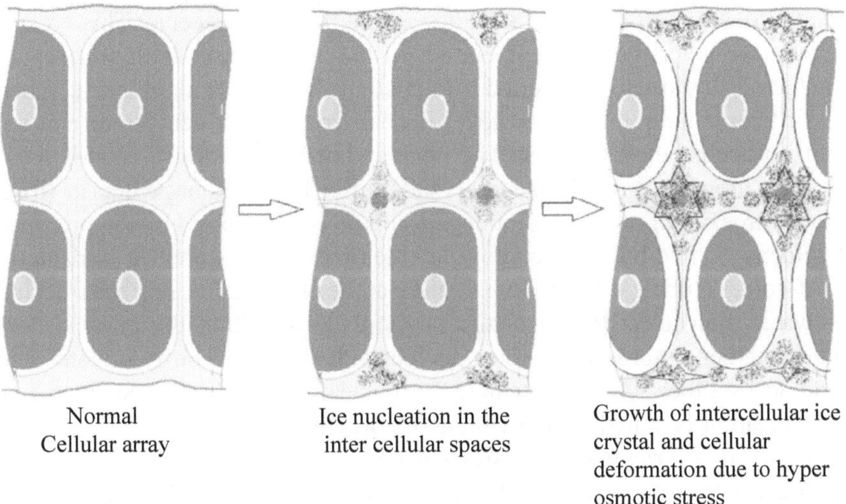

Fig. 3: Pictorial representation of intercellular freezing and hyper-osmotic plasmolysis during a frost induced freezing event

Direct freeze damage occurs when ice crystals form inside the protoplasm of cells, it is known as **intra-cellular freezing** and it is not very common under natural conditions [11]. But, under controlled environment conditions it was observed that some of the subtropical species like papaya, banana, jackfruit etc. under low humidity conditions do experience intracellular freezing together with intercellular freezing when these plants were shifted from room temperature to a rapid freezing environment (<-4^{0}C) having temperature drop rate of 4^{0}C per hour. Earlier, Siminovitch *et al.*[13] also reported cell death in winter rye plants due to freezing of intracellular supercooled water when these plants were cooled at a temperature drop rate of 8^{0}C per minute up to -12^{0}C. Also, the tremendous growth of intracellular ice crystal in the intra-cellular space goes beyond the elastic limits of the cellular membranes and thus leads to their rupturing causing irreparable damage to the cells. Therefore, the cells experiencing inter-cellular ice formation may be reverted to normal but the cells which experience rupturing of cellular membranes due to intra-cellular freezing can never be brought back to their normal condition.

Frost and freeze injuries are thus closely related and the damage looks the same. In both the cases the ice crystal forms in the water-filled plant tissues, dehydrating cells and disrupting membranes. The affected cells collapse and appear dark after some time. Frost damage occurs mostly during a radiation cooling event whereas freeze damage is common under advection/convection freeze event. Radiation freeze occurs on clear, calm nights when plants radiate (loose) more heat into the atmosphere than they receive. This creates temperature inversion (the temperature of the air increases with altitude) in which cold air close to the ground is trapped by the warmer air above it. When the air temperature at the plant level is near freezing, the temperature at the plant surface is below freezing. The plant and plant parts which cannot avoid ice nucleation and growth of ice crystal, suffer heavy damage due to frost-induced freezing. Freezing-point depression of 1-2^{0}C, caused by the presence of solutes [3,14] and by super-cooling, is often too slight to prevent freezing in moist-colder climates. As, the temperature at the plant surface reaches freezing the water molecules present at the phylloplane or within the plant system come together to form a stable ice nucleus either spontaneously or catalyzed to do so by other substances called ice nucleating agents. The spontaneous ice nucleation is called **homogeneous ice nucleation** whereas the catalyzed one is called **heterogeneous ice nucleation**. The homogeneous ice nucleation of pure water takes place at -38.5^{0}C or below, at this temperature even the finest droplet of water also freezes [14]. Even for single cells, if they have not frozen or been freeze dehydrated at a higher temperature, then, they will freeze internally near -40^{0}C (the homogeneous nucleation point being depressed by the solutes). However, a cell with highly viscous contents, such as any cells dehydrated by the growth of extracellular ice, are likely exceptions. These cells may form a glass (Vitrify) rather than freeze [13], and this may explain why some tree species can survive temperatures down to liquid nitrogen temperature[6, 7]. In nutshell, different mechanisms which result in plant damage due to frost induced freezing stress are illustrated there in Figure 4. The illustration depicts that the frost induced freezing affects plants directly and/or through soil heaving (freezing of soil water resulting in the upward eruption of soil profile). The freezing induces intercellular ice formation in plant tissues experiencing this stress. This further induces osmotic stress leading to cellular dehydration. It affects the cellular lipid–water balance as well as the protein-lipid-protein structure of the cell wall. This intercellular freezing may progress further to intracellular ice formation which further results in complex cellular disorders like denaturation of enzymes & proteins, membrane damages, cellular electrolyte leakage etc. and may lead to increased ROS production, organelles dysfunctions, cessation of biological activities and complete cellular disruption. Such types of damages are irreparable and result in the death of an organ or the whole plant. On the other hand, the freezing of soil water either directly induces freeze damage to the plant roots results in soil heaving that causes damage to the fine roots of the plants or results in the uprooting of small plants or plants with the shallow root system. The freezing of soil water creates an acute water deficit leading to the cessation of water & nutrient uptake processes. Such type of freezing stress also leads to xylem embolism resulting in the death of the plants.

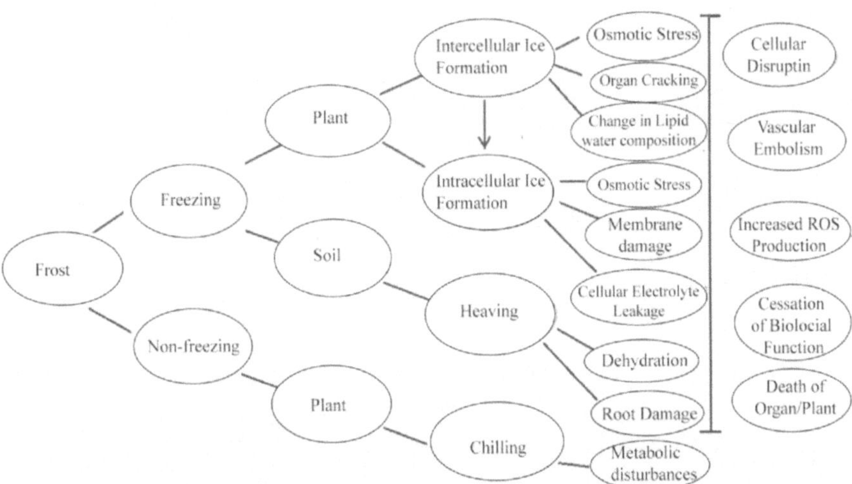

Fig. 4: Illustration of different types of frost induced freeze damage in subtropical fruit plants

The non-freezing influences of frost stress cause metabolic disturbances in subtropical fruit plants and affect the processes associated with acclimation of the plants to low temperature stress. Such plants often oscillate between inductive periods of active and non-active growth and suffer heavily if get exposed to freezing stress.

5

Biomolecular Basis of Cold Stress Tolerance and Acclimation

During winters, many of the evergreen species growing in the subtropics suffer chilling or freezing damage and the extent of damage however, depends greatly on their ability to perceive the low temperature stimulus and the capacity to modify the physiological or biochemical functioning at the molecular level; this phenomenon has been named as **cold acclimation** [15,16]. The process of cold acclimation is more complex in evergreen broad-leaved species; in these species, the long-lived leaves allow the plants to take advantage of every favourable opportunity for dry matter production. However, this leaf longevity requires various protection mechanisms to survive the dangers of low temperature and freezing stress in the winters.

A wide range of cellular mechanisms in subtropical plants gets affected by low-temperature depending on the intensity and duration of stress. Major structural or molecular changes which occur in the plant system upon exposure to the low temperature stress are:

- Changes in membrane structure and lipid composition of the cell [17, 18].
- Leakage of cellular electrolytes and amino acids.
- Shift in the electron flow toward alternate pathways [19].
- Alterations in protoplasmic streaming and redistribution of intracellular calcium ions [20].
- Modifications in protein metabolism and enzyme activities [21].
- Alterations in a number of ultrastructural cellular constituents; such as plastids, thylakoid membranes and phosphorylation of thylakoid proteins and mitochondria [22].
- Transitory changes may also be observed due to short-duration exposure to low-temperature stress however, prolonged exposure causes propagation of ice throughout the vascular system of the plant leading to plant necrosis or death.

The onset of these changes starts as soon as the plants sense the declining trend of temperature.

Sensing of cold stress in plants and signalling mechanisms

Cold stress sensing by the plants is still a complex to be understood completely. A detailed account of the review on the role of biomolecules in cold sensing and subsequent physiological modification in response to cold stress has been presented by Sharma and Kumar [23]. Various studies conducted so far have inferred that a specific sensor for low-temperature stress has not been identified, yet; however, multiple primary sensors are thought to be involved in this process and each sensor is believed to perceive a specific aspect of the stress. These sensors are believed to have a role in a distinct part of the cold signalling pathway. The primary site of cold stress sensing in plants could be associated with membrane fluidity, protein and nucleic acid conformation, metabolite concentration, and cellular redox status etc.[24]. Further transportation of cold stress signal seems to be localized to plasma membrane or membrane of the nuclear envelope which has an association with the regulation of cold-inducible Ca^{2+} transients. The cold stress-induced Ca^{2+} signature has been elucidated and decoded by different pathways. CaM (calmodulin) and CMLs (CaM-like), CDPKs (Ca^{2+}-dependent protein kinases), CCaMK (Ca^{2+}-and Ca^{2+}/CaM-dependent protein kinase), CAMTA (CaM-binding transcription activator), CBLs (calcineurin B-like proteins), and CIPKs (calcineurin B-like proteins) and CIPKs (CBL-interacting protein kinases) are some of the groups of Ca^{2+} sensors found in plants. CDPKs are positive regulators of gene expression for cold tolerance in plants, according to genetic studies [25] however, calmodulin-3 has been reported to be a negative regulator of gene expression and cold tolerance in plants [26]. CBLs interact with and regulate the CIPKs to relay the Ca^{2+} signal. The CBL1 mutant has a chilling sensitive phenotype and interacts with CIPK7 to modulate cold sensitivity [27]. Through binding to a regulatory element (CG-1 element, vCGCGb), CAMTA3 has been identified as a positive regulator of CBF2/DREB1C expression [28]. Plants with the CAMTA2 and CAMTA3 double mutants have been found to be sensitive to freezing temperatures.

In addition to the plasma membrane, the chloroplast is also involved in temperature sensing. Excess photosystem II (PSII) excitation pressure caused due to an imbalance between the capacity to harvest light energy and the capability to dissipate this energy through metabolic activity at low temperatures leads to formation of reactive oxygen species (ROS). Protein phosphorylation in response to cold and the suppression of protein phosphatase activity also provide a mean for the plant to sense low temperature. Cold signalling and cold tolerance are both regulated by the MPK (mitogen-activated protein kinase) cascade. It has therefore been speculated that environmental stresses, including cold stress are first perceived by the receptors present on the membrane of the plant cells. The signal is then transduced downstream through many signalling pathways. Studies have shown that such pathways are often activated in concert, the components of which are calcium, reactive oxygen species, protein kinase, protein phosphatase and lipid signalling cascades. It is believed that specificity to a certain pathway is achieved by the combination and timing of the activation of different signalling pathways. The calcium-binding proteins sense the change in cytosolic calcium level. These proteins do not possess enzymatic activity but undergo conformation changes in a calcium dependent manner. The conformational changes in calcium-binding proteins make them interact with other proteins and to initiate a phosphorylation cascade. Through this cascade the plant cells target major stress-responsive genes or transcription factors. The transcription factors also regulate the expression and function of genes, which ultimately leads to plant acclimation and survival during unfavourable conditions [29,30,31,32]. Individual plant cell responds to the environmental stress in this way and then the whole plant acts synergistically. The change in the gene expression governed by the signal cascade mechanism also induces a change in the genes participating in the formation of plant hormones such as abscissic acid, salicylic acid and ethylene. These hormones may amplify the same cascade or may initiate a new signalling pathway. Additionally, several other cellular components are also involved in the stress signal transduction mechanism. These accessory molecules may not directly participate in signalling but participate in the modification or assembly of major signalling components. Mainly, such components are protein modifiers and act in post-translational modification of signalling proteins. Such modifiers are involved in myristoylation, glycosylation, methylation and ubiquitination of signalling proteins [31,32].

During cold stress, lipid molecules play a vital role in signal transmission. As lipid signalling has received less attention, phosphatidic acid, which is generated by phospholipase-D and concerted action of both phospholipase-C and diacylglycerol kinase, has been proposed as a secondary messenger molecule present in the cellular membranes. Phosphatidic acid is produced quickly and transiently in response to a variety of stress stimuli and has been proposed to function as a second messenger [33]. Under normal growing conditions, phosphatidic acid represents a small percentage of membrane lipids, but its levels increase significantly in response to a wide range of stresses, including cold stress [33,34]. Phosphatidic acid can act in signal transduction through a variety of mechanisms. Its primary functions revolve around the engagement and control of target proteins' enzymatic activity. Phosphatidic acid, on the other hand, could affect the enzyme activity of proteins already present in the recruited proteins or protein complexes. A number of phosphatidic acid binding proteins have been discovered in a variety of organisms, and several have been postulated to play a role in low temperature stress signalling pathways [35].

Abscissic acid is an important phytohormone that plays a crucial role in several plant stress responses, including cold stress [32,36]. Furthermore, phospholipase-D has been linked to reactive oxygen species, which are known to be involved in abscissic acid and in cold stress responses [32,37]. As a whole, it seems that phospholipases are involved in multiple aspects of both the overlapping and distinct signalling networks that are activated by cold stress. It has been suggested that ethylene receptors (ETH) are positive regulators of freezing tolerance. One possibility is that the negative regulatory role of cytokinin (CK) signalling in freezing tolerance is at least partially dependent on ABA signalling [38]. This study expands our knowledge of the network of ETH and CK signalling in plant responses to environmental stresses.

Lipids and Membrane Fluidity

One of the common mechanism plants use to deal with cold stress is to change its membrane lipid composition for better stability and integrity of the membrane at low-temperature stress. The plasma membrane is mainly composed of phospholipids and proteins. These lipids in the plasma membrane are either unsaturated or saturated fatty acids. Membranes containing saturated fatty acids solidify faster at a temperature higher than those containing more unsaturated fatty acids. Thus, the relative proportion of the two types of fatty acids in the lipids of the plasma membrane determines the fluidity of the membrane [39]. It is now known that one of the membrane lipid response actions, when exposed to low-temperature, would be to increase the unsaturation of fatty acids [40,41]. However, this response differs substantially from plant to plant and even the same plant can act differently depending on the way of exposure to low-temperature stress *i.e.* acclimated (3-4°C) versus non-acclimated plant (22-24°C) [42]. In low temperature tolerant species, the unsaturated fatty acids are synthesized from saturated fatty acids by fatty acid desaturases, which convert single bonds to double bond [43]. The extent of desaturation is largely regulated by the genetic and environmental factors. Temperature is a significant environmental component governing the level of desaturation of individual fatty acids [44,45].

Role of reactive oxygen species (ROS)

Another stress response in the plants is the stimulated production of reactive oxygen species (ROS) like OH·, O_2·, H_2O_2 etc. These species cause considerable damage through the peroxidation of membrane lipid components and also through direct interaction with various macromolecules. Cells adopt different mechanisms to keep the ROS level in check. However, low ROS concentration participates in the signal transduction mechanism. These ROS are scavenged by low molecular weight antioxidative metabolites like glutathione, ascorbic acid, α-tocopherol and antioxidative enzymes like; catalase, ascorbate peroxidase and superoxide dismutase. However, under stress conditions, the production of free radicals exceeds the overall cellular antioxidant potential that leads to oxidative stress, thereby adversely affecting plant growth. One of the consequences of uncontrolled oxidative stress is; damage to cells, tissues and organs caused by oxidative damage. The role of ROS has been reported to be in processes leading to plant stress acclimation [46]. This finding indicates that ROS are not simply toxic by-products of metabolism but also act as the signalling molecules that modulate the expression of various genes, including those encoding antioxidant enzymes and modulators of H_2O_2 production [46,47]. It has long been recognized that excessive levels of free radicals or reactive oxygen species (ROS) can inflict direct damage to lipids. The two most prevalent ROS which affect the lipids are mainly hydroxyl (HO·) and hydroperoxyl (HO_2.) radicals.

The hydroxyl radical (HO·) can be produced from O_2 in cell metabolism and under a variety of stress conditions. A cell produces about 50 hydroxyl radicals per second, about 4 million free hydroxyl radicals are produced in a day which can neutralize or attack biological molecules [48]. The H_2O· in biological systems is formed through a redox cycle through the Fenton and/or Haber-Weiss reactions, wherein the free iron (Fe_2^+) reacts with hydrogen peroxide (H_2O_2) to produce Fe_2^+. The HO_2· is a much stronger oxidant than the superoxide anion radical and could initiate the chain oxidation of polyunsaturated phospholipids, thus leading to impairment of membrane function [49,50,51]. Lipid peroxidation can generally be described as a process in which oxidants such as free radicals or non-radical substances attack lipids containing carbon-carbon double bonds, especially polyunsaturated fatty acids (PUFA), which involve the extraction of hydrogen from carbon and insert oxygen to generate lipid peroxy radicals and hydroperoxides [52].

Glycolipids (GL), phospholipids (PL), and cholesterol (Ch) are also well-known harmful and potentially fatal targets of peroxidation modification. Lipids can also be oxidized by enzymes such as lipoxygenase, cyclo-oxygenase, and cytochrome P_{450}. In response to membrane lipid peroxidation, cells promote survival or induce cell death according to specific cell metabolism and repair capabilities. Under physiological or low-lipid peroxidation conditions (subtoxic conditions); cells stimulate their maintenance and survival through constitutive antioxidant defense systems or activation of signaling pathways that regulate antioxidant proteins, leading to adaptive responses to stress. In contrast, under medium and high lipid peroxidation rates, the extent of oxidative damage exceeds the repair capacity and cells induce apoptosis or programmed cell death through necrosis; both processes will eventually lead to molecular cell damage, thereby promote the development of various diseases. Once lipid peroxidation begins, chain reactions will occur until the termination product is produced. In addition, the low temperature induces harmful effects in lipid composition of bio-membranes, additional factors like synthesis and accumulation of compatible solutes and cold acclimation induced proteins changes in carbohydrate metabolism [53,54,55] and boosting of radical scavenging potential of cell etc. [56,57] also contribute to the damage. Thus, cold stress results in loss of membrane integrity leading to solute leakage; it disrupts the integrity of intracellular organelles causing to loss of compartmentalization and impairment of photosynthesis, proteins assembly and general metabolic processes.

Physiological functions and acclimation

As discussed above, the primary mechanism involved in cold acclimation is related to several processes like molecular and physiological modifications occurring in plant membranes. Under exposure to low temperature stress, at physiological level, the photosynthesis is strongly affected. Further, the growth arrest caused by cold stress reduces energy utilization and leads to feedback inhibition of photosynthesis [21]. In cold acclimated plants, photosynthetic activity is maintained by increasing the abundance and activity of several Calvin Cycle enzymes [58]. This recovery is related to the increase in the level of thylakoid plastoquinone and the corresponding increase in the apparent size of the electron donor pool to PS-I system [59]. Consequently, non-photochemical quenching increases in cold-stressed leaves in parallel with increased zeaxanthin levels to compensate for the reduced electron consumption by photosynthesis. Zeaxanthin protects the PS-II reaction centres from over-excitation [60]. However, Ruell and Zachowski [61] found that energy dissipation through non-photochemical cooling and electron transport is not only improved after cold acclimation, but also helps prevent oxidative damage. Xanthophylls are not photosynthetic pigments, but these xanthophylls (especially violaxanthin, anthocyanin and zeaxanthin) help protect the photosystem and therefore, their abundance increases at low temperatures [62]. Xanthophylls act as a natural antioxidant and may cause damage to chloroplasts [63,64]. Besides, in response to cold stress and other osmotic stresses, plants accumulate a variety of compatible solutes, including free sterols, sterol glycosides and acylated sterols, glycosides, raffinose, arabinoxylan and soluble sugars. Plants also accumulate other solutes, such as glutamic acid, amino acids (alanine, glycine, proline, and serine), polyamines, and betaine in response to cold stress [65,66]. These different molecules usually degrade after overcoming pressure and therefore called osmoprotectants or compatible solutes.

Carbohydrate and cold stress

According to reports, the instantaneous sensitivity of carbohydrate metabolism to low temperature is higher than other components of photosynthesis [67]. Although the exact role of soluble sugars has not been determined, their accumulation in cold acclimated plants suggests that they might be having osmo-regulatory role while acting as cryoprotectants or signaling molecules [68]. Sugars play multiple roles in low temperature resistance; as typical compatible penetrants, they help maintain water in plant cells, thereby reducing the availability of water to the newly formed ice nuclei in apoplasts [61]. Sugars also protect plant cell membranes during cold-induced dehydration, replacing water molecules in establishing hydrogen bonds with lipid molecules [61,69]. Accumulating pieces of evidence suggest that sugar functions as an osmotic substance, a membrane stabilizer and might be an antioxidant in cold stress responses [70,71,72]. The metabolism of sugar is a dynamic process regulated by numerous enzymes whose expression and activities change in response to low temperature stimulation [73]. Moreover, carbohydrates may also act as scavengers of ROS and contribute to increased membrane stabilization [74]. Additionally, sugars exhibit flexible homeostasis between source-and-sink tissues or organs due to cell-to-cell and long-distance transport mediated by various sugar transporters [75]. Various forms of soluble sugar participate in the physiological response of cold stress, depending upon the type of plant species. For example, treating rice seedlings with fructose or glucose before low temperature exposure improves their cold resistance. Cotton cotyledon discs floating on a sucrose solution in the dark were less injured by cold than those on non-sugar solutions [76]. The oligosaccharides, raffinose and stachyose are especially associated with cold hardiness, low temperature and dormancy [76,77]. Moreover, the concentration of sucrose, the most easily detectable sugar in cold-tolerant species, increases many fold during exposure to low temperature [78]. Although elevated trehalose levels are related to the tolerance of transgenic plants expressing heterologous microbial genes to abiotic stress, the role of endogenous trehalose in higher plants remains unclear. This sugar has a unique reversible water absorption capacity and appears to be superior to other sugars in protecting biomolecules from damage caused by drying [79]. During the exposure to low-temperature, the starch content typically declines following hydrolysis, and there is a corresponding increase in the concentration of free saccharides [80,81]. However, in many cases, increase in the levels of both soluble sugars and starch have been noticed during cold acclimation.

Osmolytes other than sugars involved in cold acclimation

a) **Proline:** This proteinogenic amino acid, in addition to being incorporated into protein synthesis during translation also has many other regulatory effects on plant metabolism. It acts as a permeate for osmotic regulation and helps stabilize the cell structure (membrane, protein, etc.), eliminates free radicals and buffers the cellular redox potential under stress conditions. It may also act as protein compatible hydrotrope alleviating cytoplasmic acidosis and maintaining appropriate $NADP^+$/NADPH ratios compatible with metabolism. In many plant species, the proline accumulation under stress has been correlated with stress tolerance and generally found to be higher in tolerant plants than the sensitive ones. Its build up occurs usually in the cytoplasm, where it acts as molecular chaperons, preserving protein structure and buffering cytosolic pH and maintaining cell redox state. It has also been suggested that its build up could be a stress signal that influences adaptive responses. The positive correlation between the accumulation of endogenous proline and improved cold tolerance has been found mostly in low temperature-insensitive plants such as barley, rye, winter wheat, grapevine, potato, chickpea, and *Arabidopsis thaliana* [82,83,84]. Therefore, it appears that proline possesses the potential to alleviate low temperature injury in chilling-sensitive plants, but for some reasons, this system fails under natural conditions.

b) **Glycine Betaine:** It is an amino acid derivative that has a role in plant osmolyte regulation under stress conditions. The accumulation of glycine betaine (GB) usually correlates with the plant's level of stress tolerance. The tolerance of plants to abiotic stress is enhanced in both types of plants that are either genetically engineered for biosynthesis of GB or being exogenously treated with GB [85]. Possibly the GB is involved in stabilization of transcriptional and translational apparatus under low temperature stress. It also stabilizes protein complexes, cellular membranes and also indirectly induces H_2O_2-mediated signalling pathways. Plants adjusted to low temperature conditions store molecules of freezing protection relevance [86]. Glycine betaine is one such cryoprotective solute that protects the activities of enzymes and proteins, stabilizes membranes and photosynthetic apparatus under chilling and freezing temperatures [87,88,89]. It too decreases the peroxidation of membrane lipids and ensures electron transport by means of complex-II in mitochondria [90]. It also reduces the peroxidation of membrane lipids and protects electron transport via complex-II in mitochondria [90]. Exogenously applied glycine betaine or the expression of transformed genes for endogenous glycine betaine synthesis increases cold tolerance in glycine betaine non-accumulating plants [91].

Lignin and cold acclimation

It is well-known that lignin fills the spaces within the cell wall to reduce water permeability and increase its stiffness. Extensive cellular studies have shown that freezing tolerance is directly associated with properties of cell wall and its permeability which particularly depends on the lignin content. This property empowers the plant cells to outpour the water and let the freezing happen in the extracellular spaces without harming the cellular integrity. The cell wall is thus a network made up of cellulose, hemicelluloses and lignins outside the cellular components and provides unique versatile solidarity to the cell under the adverse environmental conditions [92,93,94]. The cell wall integrity and structure are actively controlled by plant developmental processes and are capable of being redesigned in response to numerous environmental stresses [95,96,97,98]. The finer adjustments in regulation of comparative amount and proportion of the constituents of cell wall structure determine the nature and function of cell wall under the conditions of stress. Remarkably, deposition of lignin phenylpropanoid polymers, which is a highly hydrophobic component in the cell wall, determines cell wall stiffness and permeability to water [99,100,101]. Research has shown that the expression of genes related to cell wall biosynthesis and redesigning gets dramatically altered under cold treatment [102]. Cryo-scanning electron microscopy (cryoSEM) revealed that both cell membrane and cell wall properties play equally important roles in cold acclimation and freezing tolerance [103].

Lignin is a major component of the plant secondary cell wall, and the amounts of lignin are altered after cold treatment in various species [104,105]. In the past decades, some genes that regulate lignin biosynthesis have been identified. One of them is the phenyl-alanine ammonialyase 1–4 (PAL) which code the enzymes that turn the primary step within the phenylpropanoid pathway to regulate the synthesis of lignin and secondary metabolites (e.g. flavonoids and salicylic acid) in the thale cress plant [106,107,108]. The blue copper-binding gene

(BCB) present in this plant is another positive regulator of lignin synthesis [109]. Recently, Hangtao *et al.* [110] reported that specifically cold-induced nuclear protein, tolerant to chilling and freezing (TCF1), interacts with chromatin containing a target gene - blue copper-binding (BCB) protein, encoding a glycosyl-phosphatidylinositol anchored protein that regulates lignin biosynthesis. These shreds of evidence show that decline in lignin content significantly improves the ability of the plant to withstand phase change of water. During this process the necessary lignin maintenance for cold tolerance is regulated by TCF mediate signaling mechanism. This finding provides the first direct molecular evidence that freezing tolerance is directly related to cell wall properties during cold acclimation.

From the preceding information concerning the role of biomolecules in cold sensing and tolerance, it can be inferred that plants exposed to freezing temperature typically show common responses like cellular dehydration and elevated oxidative stress. However, the extent of damage caused due to this stress depends greatly on the biomolecular constitution and physiological functioning of the plants. Numerous research findings support the notion that Ca^{2+} evoked stiffening of cell wall will cause active cytoskeletal transcription that helps the plants in developing adaptive capacity against low-temperature stress. Further, the increase in ROS under low-temperature stress and their linkage with lipid peroxidation has proven to cause damage under cold stress conditions. Therefore, the strategies resisting ROS induced lipid peroxidation can serve as a key mechanism for the protection of plants against frost-induced freezing damage. The role of carbohydrates is also important as trehalose sugars have the unique capacity for reversible water absorption and appear to be superior to other sugars in protecting plant biomolecules from freeze desiccation damage. Proline and glycine-betaine are the primary osmoregulators in the plant system under stress condition. There exist a higher level of these osmolytes in cold-tolerant species and this level gets enhanced in the case of acclimated plants. The exogenous application of these amino acid groups has been speculated to be beneficial for improving cold tolerance in non-accumulating plants. Elaborated findings on TCF mediated signaling of lignin biosynthesis for modifying cell membrane and cell wall properties and over-expression of CBF genes resulting in enhanced freezing tolerance needs to be worked out beyond *Arabidopsis,* for extending benefits to the commercial crops. Therefore, more focused research on the development of plants that possess genes for the accumulation/synthesis of beneficial biomolecules is needed. A well-organized approach is required to combine and investigate the molecular, physiological, and metabolic aspects of low-temperature stress tolerance both at the cellular and whole plant level together with consideration for varied agro-ecological situations of the frost prone areas.

6

Frost Sensitivity of Horticultural Crops

All the plant biochemical processes are sensitive to change in temperature. Freeze or low-temperature stress in plants becomes lethal when the temperature is lowered beyond a critical threshold for a sufficient duration to cause irreversible damage to plant growth and development. This stress influences countless biochemical reactions which have certain temperature optima for their proper functioning. Deviations both on the upper or lower limits of the temperature optima are critical for the biological activity of the plants. Schematic illustration of the influence of temperature on some of the important plant processes and growth rate are presented here in Figure 5. The responses of various life processes and growth to the change in temperature are rarely linear, observe an exponential path *w.r.t.* change in temperature [111,112,113]. Apart from the influence on physiological processes, low-temperature stress usually creates an imbalance between light absorption and light utilization by inhibiting the activity of the Calvin-Benson cycle. Enhanced photosynthetic electron flux to O_2 and over reduction of the respiratory electron transport chain results in reactive oxygen species (ROS) accumulation during the low temperature and this causes an oxidative stress [114]. Tolerant plants have evolved a variety of defensive responses to these temperature changes and are capable of minimizing damage to ensure the maintenance of cellular homeostasis [115]. But, the non-tolerant ones; especially, the tropical & subtropical evergreen species lack such type of defence mechanisms and suffer heavy damages due to low-temperature stress.

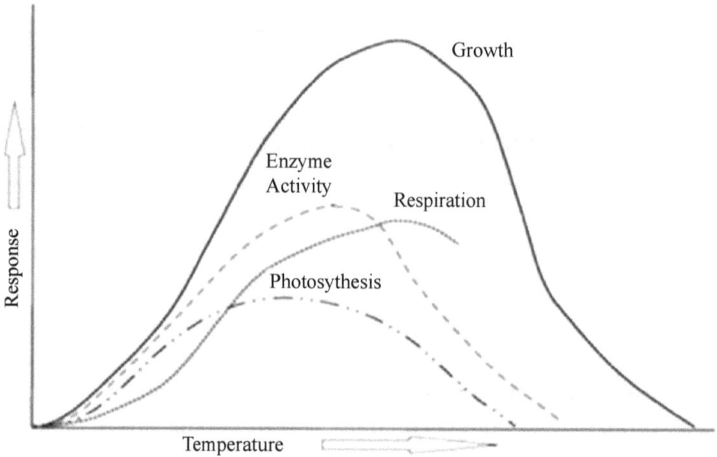

Fig. 5: Schematic representation of effect of temperature on vital plant growth processes. Source: Fitter & Hay[111], Zrobek-sokolink[112], Hasanuzzman[113]

Table 1: Optimum temperature ranges for growth of some sub-tropical fruit trees

Fruit Tree Species	Optimum Temperature (°C)	Reference
Mango	24-27	Davanport [116]
Litchi	25-35	Tindall [117]
Guava	23-28	Verheij [118]
Banana	20-35	Viktorova [119]
Longan	20-25	Verheij [118]
Grapes	10-35	Larcher [120]
Jackfruit	16-28	Haq [121]

The mean temperature range for optimum growth of most tropical fruit plants is about 24-30°C [122,123]. However, the optimum temperature range for better growth of mango, litchi, guava, banana, longan, grapes and jackfruit etc. is presented in Table 1 depicts that each of these fruit species has different ranges of optimum temperature for their peak performance. But, it is also a fact that these species have adapted to a certain extent beyond their original habitat, especially in the subtropics. And, due to difference in the fruit maturity time from the tropics or their original habitat, the area under these fruits is under expansion in the subtropical regions. In the subtropics, the growers derive higher revenue from the produce which is harvested at such a time when it is over from the tropics. But, under these growing conditions, due to deviation in the cardinal temperature requirements of these crops; every now and then the plantations of these fruit crops suffer heavy damage due to frost and freezing temperatures. Sometimes, the loss is so high that some of the crop plantations get devastated completely during severe frosty winters.

For proofing the performance of tropical and subtropical fruit species against frost in the sub-optimal growing conditions of the subtropics, it is imperative to devise an effective mechanism for dealing with frost and low-temperature stress. And, for devising appropriate protection mechanisms, knowledge of critical temperature at which irreversible damage occurs to different species is very essential, it can serve as a pivotal factor in the process of frost protection decision making. Also, it helps in devising the strategy for the prevention or delaying of ice nucleation so as to protect the plants from frost-induced freezing damage. There are some studies on critical damage temperatures for some crops, most of these are based on the range method for defining low and high limit of critical temperature. But, the information on these temperature ranges is needed to be used carefully while setting the 'On' and 'Off' timing of a frost protection device, a slight variation from the critical temperature range can cause huge loss to the plantations under severe frost conditions. Usually, the researchers use long-term commercial damage records with temperature measurements from standard meteorological laboratory shelters. The factors (like temperature sensor, shielding, mounting height, etc.) which influence the actual reading of meteorological instruments are sometimes not reported. Also, there exists some influence of microclimate on the observed values of temperature, may be differences of 1.0 °C or more within a few meters in an orchard during a frosty night, measured at the same height above the ground on flat terrain. Therefore, it is doubtful that critical damage temperature values from meteorological shelter instruments are to be accepted universally or not. On the other hand, many researchers use small pieces of cut branches from trees and place them in climate control chambers where these excised branches are cooled to a range of subzero temperatures and the observations on critical damage temperature and level of damage are taken. Such type of experimental conditions and observations are quite different from field measurements, the microclimate inside of a climate control chamber is not the same as that of branches exposed to the sky under open field conditions. For example, one could determine the amount of damage for branches exposed to 30 minutes at a range of temperatures, but under natural field conditions, the uncut branches within a tree will have a broad range of temperatures. Branches at the upper tree canopy which are exposed to the sky will probably be colder than the air temperature. Conversely, branches in the central canopy are likely to be warmer and thus less prone to damage. In case of spring frost, in deciduous trees at the time of leaf emergence, the coldest temperatures are near the ground level due to temperature inversion but, a few days later when trees have most of their leaves fully expanded, on a radiation frost night, due to decreased temperature inversion rates, the minimum temperature occurs there near the canopy where most leaves and flowers or fruit-lets are growing. Therefore, the temperatures from the weather shelter only provide a rough estimate for the expected level of frost threat.

For drawing field replicable findings, it is imperative to run the experiments carefully in the controlled freezers with slow air movement, while working on critical temperature estimation. The air temperature in the freezer should be lowered in small pre-determined steps and held for about 30 minutes or more after each step to allow the buds to come into equilibrium with the surroundings. It is generally misunderstood that buds are required to be kept for 30 minutes or so for observing the damage; in actual the short periods (2 to 4 hours) of low temperature exposure are less important than how low the temperature goes [128]. But, in practices, it has been observed that the duration of low temperature exposure does matter in such studies. The damage has been observed only in those cases where exposure was sufficient for ice nucleation and propagation of ice crystal within the plant system. The extent of low temperature also matters equally in causing the damage. Under natural convective cooling conditions also, the duration of time for which the low temperature persists determines the level of damage to the plants. Under radiative cooling the plant tissues cool at a rate dependent on the radiation balance and the temperature difference between the tissue and its environment. Therefore, if the air temperature suddenly drops by several degrees the tissue will rapidly cool below critical levels and it results in freeze injury. Under sustained low-temperature stress if the plant tissue contains super-cooled cellular fluid, a small mechanical shock or fluttering of the leaves and buds by wind or other sources could initiate ice crystal formation resulting in damage even if the tissues are above the growth chamber-determined critical temperature values. However, the chamber values provide guidelines as when freeze protection measures need to be implemented.

To simulate the natural frost-induced freezing conditions, Sharma [124] conducted studies in customized walk-in frost chambers for defining the critical temperature of damage to the subtropical fruit tree species under low-under low-temperature stress. **Critical damage temperature-LT_{50}**, the temperature at which 50% of the maximum electrical conductivity of the leaked cellular contents of a species is observed, was defined for some of the subtropical plants viz. galgal, guava, jackfruit, jamun, karonda, lime, litchi, loquat, mango and papaya. Three years old container-grown plants were subjected to normal pre-winter acclimation up to December 15. These plants were then subjected to different freezing temperatures at a step-down trend of 0.5^{0}C starting from 2.0^{0}C (which is the usual minimum air temperature prevailing by December 15). One step of temperature was maintained for one hour. The procedure described by Soliemani *et al.* [125] and modified by Sharma [126,127] for electrolyte leakage studies is presented here under:

After each step of freezing temperature exposure, excise two leaves from the top canopy and make small chips, each measuring 1cm^2, with cutting block or knife immediately after excision from the plant. Put 25 chips in 50ml of distilled water in a beaker and seal the beaker immediately with plastic film to avoid evaporation of water. Put the beaker on a shaker for 6 hours and then remove the chips with a plucker and take the Electrical Conductivity of the water. Run a similar sample of the species under study after complete freezing exposure. Compare the two conductivities. This method of estimating the frost damage is called ***Relative Electrolyte Leakage (REL)*** *method. The temperature at which relative electrolyte leakage was 50% of the fully frozen leaves was defined as* LT_{50} *for that particular species.*

During these studies, the initiation of the freezing event and the subsequent spread of freeze in the leaf tissues were monitored through **Infrared Video Thermography (IRVT).** The observations on LT_{50} of some of the species studied are presented hereunder in Table 2:

Table 2: Critical temperature (LT_{50}) for frost-induced freezing in subtropical fruit species

Subtropical fruit specie	Galgal	Guava	Jackfruit	Jamun	Karonda	Lime	Litchi	Loquat	Mango	Papaya
LT_{50} (°C)	-4.0	-3.5	-1.0	-2.5	-3.5	-4.0	-3.0	-4.5	-2.5	1.5

These studies were further extended for determining the relative susceptibility of subtropical fruit crops to frost-induced freezing. The container grown plants were kept at -2°C for three hours in the walk-in growth chamber and the REL was measured as per procedure described above. The per cent value observed for the REL of different species is presented in the bar diagram presented in Figure 6. The species which were damaged more were having higher REL value and thus regarded as more susceptible to frost induced freeze damage. The laboratory observations were also validated with IRVT studies and field observations. The relative order of frost sensitivity of the subtropical fruit plant species to frost was finalized as follows:

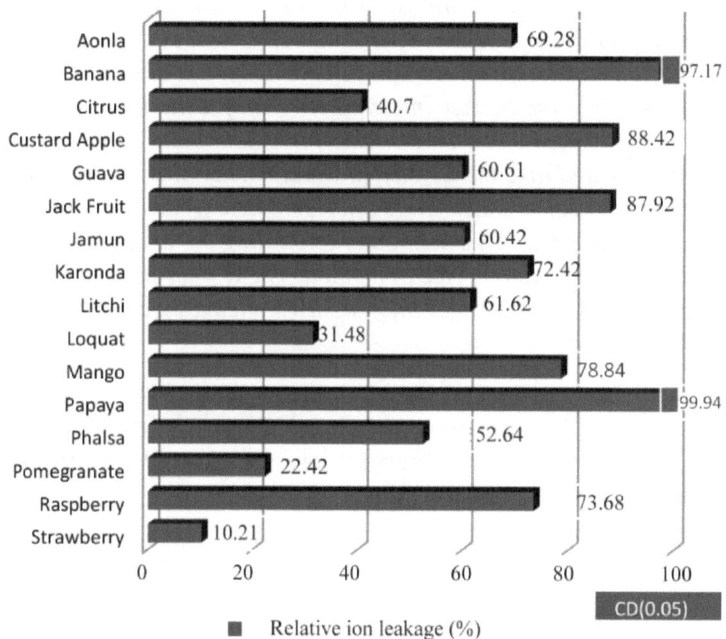

Fig. 6: Relative ion leakage (%) from leaf tissue of different subtropical fruit species exposed to freezing temperature

Increasing order of frost sensitivity of fruit crops:

Strawberry > pomegranate > loquat > citrus > phalsa > guava ≈ jamun ≈ litchi > aonla≈ karonda≈ Raspberry ≈mango ≈ jackfruit > custard apple > banana > papaya

Varietal sensitivity of Mango to frost: Varietal sensitivity of some mango varieties to frost was also worked out through a similar procedure and the increasing order of sensitivity of mango varieties was found to be as:

Increasing order of frost sensitivity of Mango varieties:

Desi> Dashehari> Langra> Fazali> Chausa> Mallika> Amrapali

The freeze sensitivity of the plants was also given by Levitt [128] who classified plants into four freeze-sensitive categories: i) tender ii) slightly hardy iii) moderately hardy and iv) very hardy. The plants which cannot avoid intercellular freezing (mostly tropical plants) were classified as tender plants; slightly hardy plants include most of the subtropical fruit trees, deciduous trees during certain periods, and fruit and vegetable horticultural crops that are sensitive to freezing down to about -5°C. In the moderately hardy plants category, the plants that accumulate sufficient solutes to resist freeze injury at temperature as low as -10°C, mainly by avoiding dehydration damage, but these are less able to tolerate lower temperatures. The plants which are capable of avoiding intracellular freezing as well as avoid damage due to cell desiccation were described as very hardy plants. These freeze sensitivity categories give general information about the level of cold stress that a plant or plant organ can endure before the occurrence of frost damage. The level of acclimation, hardening and phenological stage are also of the same importance while describing the sensitivity of the plants to freeze damage.

Another classification of the cold sensitivity of some the fruits and vegetables into most susceptible, moderately susceptible and least susceptible species was given by Wang and Wallace [129] which is presented here in Table 3.

Table 3. Susceptibility of fresh fruits and vegetables to freezing

Most Susceptible	Moderately Susceptible	Least Susceptible
Apricots, Asparagus, Avocados, Bananas, Beans, Berries (except cranberries), Cucumbers, Eggplant, Lemons, Lettuce, Limes, Okra, Peaches, Sweet Peppers, Plums, Potatoes, Summer Squash, Sweet potatoes, Tomatoes	Apples, Broccoli, Carrots, Cauliflower, Celery, Cranberries, Grapefruit, Grapes, Onion (dry), Oranges, Parsley, Pears, Peas, Radishes, Spinach, Winter Squash	Beets, Brussels sprouts, Cabbage(mature and savoury), Dates, Kale, Kohlrabi, Parsnips, Rutabagas, Turnips

Source: Wang and Wallace[129]

7

Frost Damage Symptoms

As the vulnerability of different plant species to frost and low-temperature stress is quite variable, the symptoms are also variable, therefore. The development of visual symptoms depends largely upon the plant genotype, growing environment and tissue type being exposed to low-temperature stress. Also, how long and how low the temperature goes during a freezing/ frost exposure also affects the level of damage and the subsequent development of the symptoms. Even, the plants which tolerate frost may develop visual symptoms in response to frost and low-temperature stress. Some of the commonly observed apparent symptoms of frost and low-temperature stress are described hereunder:

- **Pigmentation:** The development of foliage fall colours is a common phenomenon in the case of some deciduous plant species like apricot, plum, peaches, maples, oaks, blueberries, sourwood, sweetgums and persimmon etc. With the decrease in day length and temperature, the photosynthetic machinery of winter deciduous plant species goes into a declining phase and chlorophyll degradation occurs. With the decrease in the amount of this dominant leaf pigment, the colours of other pigments like anthocyanins, carotenes, xanthophylls etc. become apparent which otherwise remain masked by the presence of chlorophyll. The gradual decrease in temperature also results in enhancement in levels of anthocyanin or β- carotene or neoxanthine contents of leaves or branches or both of some of the low-temperature sensitive species [130]. The synthesis of these pigments gets enhanced as a measure to combat the decreasing temperature stress and this gives a yellow, red, bronze, purple or orange appearance to the senescing leaves in autumn or early winter. This phenomenon is apparent in some subtropical evergreen species also; like guava, karonda, jamun, etc., which show enhanced pigmentation mostly under chilling stress conditions rather than under freezing stress.
- **Flaccid or wilting foliage**: During periods of low temperature, leaves of some evergreen subtropical tree species like citrus give a flaccid appearance. They usually show wilting signs as a measure to combat the

low-temperature stress. Such situations generally occur when the low-temperature stress is there for the plant but it is in its tolerable range. During the daytime, when the temperature reaches normal, these leaves gradually regain their turgor and recover.

- **Browning, scorching or burning symptoms**: Under severe frosty situations when the temperature goes below freezing range and beyond the tolerable limit of the plants, there occurs permanent damage to the foliage. This type of damage does not appear immediately but, becomes visible after few days of damage. The leaves appear dark brown, scorched or burnt with brittle or curled appearance after complete death of the leaf tissues. In severely damaged leaves and leaf pedicel, the browning or darkening of conducting tissues may be noticed 2 to 3 days after the intense freezing exposure. The leaves remain attached to the shoots and don't fall after such damage. Severe frosty conditions involving rapid temperature drop results in such type of damage and development of burnt foliage symptoms in plants. This type of damage is very serious as the freeze progresses down to the plant's vascular system and adjoining buds through the leaf pedicel leading to blackening of conducting tissue which is called **Black Heart disorder**. In plants with dense canopy, these symptoms are seen only on the upper canopy of the plants; lower foliage gets shielded by the upper canopy and saved from frost damage. This type of damage reduces the photosynthesizing area of the plant and thereby increases the flux of reactive oxygen species in the plant system. Most of the subtropical fruit species suffering frost damage in winter shows this type of symptoms after complete death and drying of the leaf tissues. People sometimes confuse this type of damage symptoms with that some bacterial wilt or blight symptoms.

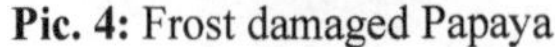

Pic. 4: Frost damaged Papaya

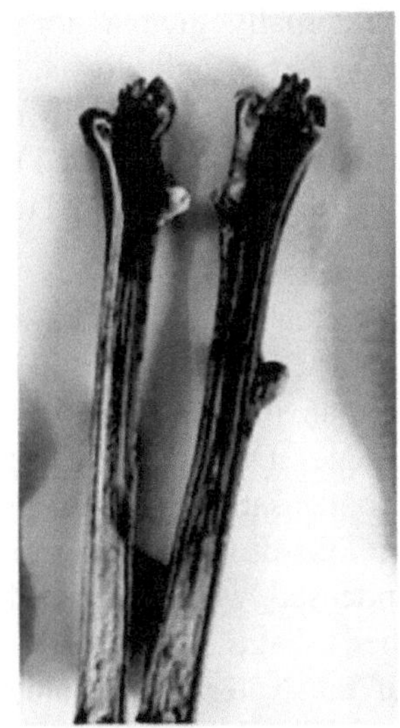

Pic. 5: Progression of frost induced freeze from bud to vaseular system in mango

- **Pale Brown patches, water-soaked lesions**: Some succulent species like *Aloe vera*, when exposed to low temperature (even above freezing) show yellowing of leaf blades and develop some water-soaked patches at various positions of the blade. Soon these affected leaf portions of such plants become brown and black. Freezing temperature exposure of such plants leads to complete disintegration of leaf tissue which gets invaded by various pathogens.

- **Withering of foliage and buds:** In the case of mature papaya and banana plants; initially, blistering symptoms appear on the foliage. If cold conditions persist, the leaf tissue withers and the leaf stalk remains attached to the main stem. After some time the stalk also abscises and the main stem left with little new growth at the top of the main stem.

  In many species like mango, litchi or other fruit species which are terminal bearer, the terminal bud shows drying up symptoms upon exposure to freezing temperature. These buds soon wither out with resumption of growth after rise in temperature. These types of symptoms appear mostly when the adjoining leaves of the bud are damaged considerably by the frost in winters. Under prolonged freezing stress, blackening of the shoots bearing such buds appears after few days.

- **Stem rot:** Very typical frost damage symptoms appear in the case of papaya and other succulent species; the main stem especially the collar region, externally give a normal appearance but, internally it gets damaged severely under low soil moisture conditions. The stem tissues near the collar region generally collapse as the temperature starts rising. The damaged cells get invaded by the rotting pathogens and lead to the fall of the trunk. The damaged trunk tissues emit a very foul smell.

- **Crotch injury**: The plants with narrow crotch angle suffer more crotch injury due to frost-induced freezing than the plants with wider crotch angle. During winters there occurs condensation of atmospheric water with the dropping temperature. The condensed water moves down from the foliage along the stem. Some of the water gets retained in the narrow crotches of the tree and serve as a source of free water for ice nucleation to occur. The ice crystal then propagates through the tender tissue in the crotch area into the plant vascular system leading to huge damage to the plants.

Pic. 6: Frost induced stem rot & defoliation of Papaya

  Also, it has been speculated by some workers that the plants with narrow crotch angle harden late or sometimes incompletely and therefore, suffer higher damage due to frost than the plants which have branches with wider crotch angle.

- **Fruit pigmentation and rots:** Frost damage symptoms on fruits are somewhat similar to the symptoms on leaves. There appears certain pigmentation on the fruit surface of some species like guava. In citrus, frost causes damage to the outer layers of fruits leading to the formation of necrotic or corky layer formation on the rind portion which is exposed to the outer atmosphere. Under extreme freezing conditions, the internal vesicles loose their integrity and become translucent. In papaya, under severe low-temperature stress there develop brown to black patches on the fruit surface. The surfaces of these patches become necrotic with disintegrated fruit tissue beneath. Under sharply cooling situations the outer surface of the fruit become lathery or corky and there occurs oozing of gum or resinous material from the corky area, whole fruit spoils soon due to invasion by the pathogens.

- **Trunk and Collar injury**: Many of the subtropical fruit species which do not possess tough bark suffer injury to their stem under freezing conditions. In species like citrus, litchi and guava; the cells of the cambium region get damaged due to frost-induced freezing. The formation of ice crystals spread fast from the cambium region to the vascular system of the plant and the whole trunk gets frozen. The wood then oxidizes and become dark (Black Heart), discoloured and vessels get filled with gummy occlusions under extremely cold conditions. A similar type of injuries occur in the collar region. The major damage to the collar region is caused due to irradiative cooling of soil near the collar region of the plant. The rapidly cooling soil draws the heat from the adjoining tree trunk and leads to a drop in temperature of the collar region beyond the tolerable limit of the plant. In the case of papaya, the plants which suffer, do not develop any apparent symptoms of injury but as they start to resume growth after winter, the damaged portion gets invaded by the rotting micro-organisms and soon the trunk collapses.

- **Trunk Craking:** Under severe frosty conditions, many woody species like Aonla, Harar etc. comparatively having tough bark also suffer damage to their trunk. The bark of these species gets cracked upon freezing temperature exposure in winters. Soon, the cracked bark tissues appear necrotic in few days. Under prolonged cold stress, severe damage occurs to the trunk tissues and their vulnerability to invasion by pathogens and insect pests increases. In some species like *Prunus* & *Cassia*, the injury so caused leads to oozing of sap and gum leading to bleeding stem disorder.

Pic. 7: Bark cracking of Aonla due to frost injury

Visualization of Non-apparent Frost Damage

The lethal frost damage to the plants become apparent usually long after the actual damage. At that time the growers are not left with any option to protect their plants. If the frost damage symptoms come to notice early, the growers can adopt some protection method to save their plantations from huge economic losses. There are a number of methods available which can be adopted for early assessment of the damage caused due to frost. Some of these methods are briefed as follows:

i) **Relative Electrolyte Leakage (REL) Method** [131,132,133]: This method is based on the measurement of electrical conductance or resistance of the solute leaked from the frost/ freeze damaged tissue. The relative measurements of the frost-damaged tissue in comparison to completely freeze damaged tissue are used for accessing the level of damage. This method doesn't give any indication about the site of ice nucleation and spread.

ii) **Vital stain Method** [134,135]: Vital stain method uses staining procedures to differentiate dead and living cells after a freezing exposure. Several stains are available but the most commonly used is a 1:1 mixture of acridine-orange and ethidium-bromide. The viable cells accumulate acridine-orange via an active uptake mechanism and show green nuclear fluorescence when viewed under ultraviolet light. The dead cells take up ethidium-bromide passively through damaged membranes and give orange fluorescence when viewed under green illumination.

iii) **Light Microscopy (LM)**[136]: It is mainly useful for detecting extracellular ice crystals.

iv) **Cryo-Scanning Electron Microscopy (Cryo-SEM)** [137]: It can detect the presence of intracellular ice, based on the presence of crystalline ice and amorphous ice inside plant cells. However, temporal imaging of ice fronts is not possible through this method.

v) **Differential Thermal Analysis (DTA)** [138,139,140]: This method measures the freezing within a plant system with the use of thermocouples which compare the output from thermocouples attached to a sample and reference. Upon freezing, the release of heat (an exotherm) occurs upon ice formation which is detected with the help of thermocouples. This method cannot detect the location at which ice is nucleated or the path of the initial spread of freezing.

vi) **Nuclear Magnetic Resonance (NMR)** [141,142,143]: It is a non-invasive method to visualize the localization of unfrozen water in plant tissues at sub-zero temperatures using NMR micro-imaging. This method clearly shows the freezing behaviour of plant tissues, mechanisms involved in the freezing behaviour are not well elaborated by this method.

vii) **X-ray Phase Contrast Imaging** [144,145]: This method uses time-resolved X-ray phase-contrast imaging for observing the growth of ice crystals in living tissue or organ for identifying the regions where the super-cooling mechanism fails and where ice crystal formation occurs in a tissue.

viii) **Infrared Video Thermography** (IRVT) [146]: This method is also based on the detection of exotherms related to ice nucleation and spread. It gives real-time imaging of plant surface temperature and helps in the detection of a location at which ice first forms and also shows the route and rate of initial growth of ice within the plant system.

ix) **Low temperature Scanning Electron Microscopy (LTSEM)** [147]: This methodology shows the position of ice relative to cells. It confirms the identity of ice by subliming it away within the magnifier and by the looks of fracture faces before and through sublimation. Freeze-substitution electron microscopy shows ice indirectly as cavities left after dissolution of ice throughout the solvent-substitution step. In freeze-fracture electron microscopy, a fracture plane passes through cell membranes detailing the freezing damage.

These methods differ greatly in spatial resolution, the decreasing order of resolution of these methods is freeze-fracture and freeze-substitution EM> LTSEM> light microscopic methods> MRI> IRVT> the unaided eye> DTA. Collectively, these methods will deliver data across a spread of scales, linking data at the whole plant and organ level to the phenomena at the tissue, cellular and sub-cellular level. Different methods possess different capacity to resolve the freezing events *w.r.t* time. The methods also differ in their ability to resolve events in time. Thus, while IRVT gives real-time images, it takes 30 minutes to obtain one MRI image; microscopic images are obtained from single time points [146]. Therefore, every method has certain advantages or limitations. Most of the above mentioned methods are costly and has a great limitation for field application.

Under natural conditions, the estimates of frost induced freeze damage are made on the basis of field surveys. The precise assessment of frost damage to the plants under natural conditions is difficult to make, it is greatly influenced by topographical, vegetation and hydrological conditions of the plantation site. From time to time many surveys are being carried out to access the level of damage and to draw strategy for protection against frost. However, most of these studies are based on isolated and somewhat subjective observations in areas with different levels of frost intensities which generally lead to contradictory findings. To address these problems which are associated with frost damage assessment, Sharma [127] devised a simple, cost-effective and highly reproducible method for visualizing the level of damage caused to the leaf tissues owing to frost-induced freezing. This method is called '**Drying**

and Rehydration Method' (DRM) for visualizing frost damage. The simple procedure of this method developed for assessment of frost damage to is as:

> *The leaf samples are collected from the affected canopy of the trees and put into a hot air oven at 50^{0}C for 24 to 48 hours for fixation of the leaf injury. After drying, the leaf samples are rehydrated by placing these in between the layers of wet filter paper for four to six hours depending upon the species (lesser time is required for rehydration of tender leaves; excessive rehydration impairs the visual clarity of the damaged portion). After rehydration, these samples were held against the light for visualizing the damaged portion. Or, these can also be visualized and photographed through an ordinary digital camera operated in the negative mode which presents the damaged area as white (more prominent than the ordinary daylight mode of photography). This photographed area can be measured with a suitable area measuring computer or mobile applications. Alternatively, the percentage of the damaged leaf area can also be calculated by plotting the entire leaf and its damaged portion on simple graph paper.*

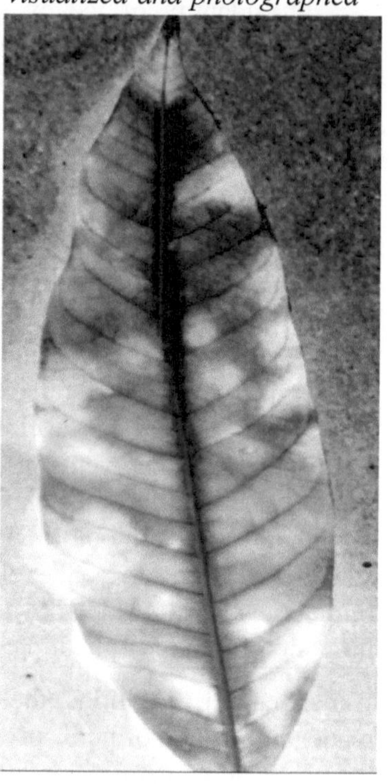

Pic. 8: Frost damaged mango leaf (image developed through drying and rehydration method)

The assessment of the damaged portion can be correlated to the frost damage incurred under the field conditions. For example, in the case of mango, it was observed that full protection against frost is essentially required in plantations where the average leaf damage area exceeds 10%. Thus, this method provides good estimates of the potential frost threat prevailing in the orchards and the level of protection needed for saving the economic losses.

8

Prediction of Frost Threat

Despite the tremendous advances in agriculture, the agrarian output is still weather and climate-dependent. Frost or low-temperature stress is a single weather-related phenomenon that causes losses greater than any other environmental or biological hazard. As the plants are immobile, they are bound to undergo this stress and suffer damage under natural condition. For avoiding and managing the adverse effects of frost-induced freezing in plants, a reliable early warning/prediction of intensity and duration of frost is necessary.

The simplest way of predicting frost stress in fruit plantations is to gauge the temperature. For the temperature information published by regional meteorological departments; the measurements are usually taken 1.25 meter above the ground to take an average of the locale which has topographical variations. The critical temperature of damage to a particular plant species is taken into consideration for deciding the level of threat. But, such types of predictions are influenced by a number of micro-climatic factors. For instance, areas which are adjacent to a large building, forest, boulders, water bodies etc. are likely to be influenced differently. Apart from these, the other factors which may influence the predictions are:

- **Clouds:** Frost is less likely on cloudy nights. A thick layer of low clouds prevents the heat radiations from Earth back into the atmosphere and it gives a warming effect in the overcast area, and frost therefore, rarely occurs on a cloudy night.
- **Wind**: Frosty conditions are affected by wind in three ways; one, it reduces the risk of frost by blowing the cool air to another place. It causes advection frost and increases the likelihood of frost. Secondly, the non-freezing air raises the temperature of the other areas where these are blown into. Thirdly, the windy situations increase the evaporative cooling and hence increase the risk of damage under freezing situations.
- **Slope:** Cold air is denser than warm air therefore, under sloppy conditions the regions lying at higher ends of the slopes escape frost damage and those lying at the basin suffer heavy damage due to frost. If the site of an orchard has a variety of slopes, a variable impact on frost proneness of the site is observed. The predictions usually do not hold good for such

situations and higher inequality between the predicted and actual values is usually observed.

- **Duration of low-temperature stress:** If the temperature drops below the critical level early after sunset, there are more chances of frost damage than under the situations where this drop occurs late during the night. The lowest temperature attained during a frosty night is always a function of the duration for which the temperature keeps on dropping.
- **Dew Point:** As, the dew point is a temperature at which there is maximum saturation of the air with moisture. When dew is formed, the heat is released within the system, keeping the air temperature at or slightly above the dew point, thus preventing frost formation. Therefore, the more is the moisture present in the air, the less likely a frost is. And, it is the reason why sprinkling of water before the expected frost helps in the reduction of damage to the vulnerable vegetation. The plants transpire a lot of water which is released into the air and increase its relative humidity. So, the plantations adjoining forests usually suffer less damage due to the higher moisture level in the air.

To forecast the extent of frost threat, many models and indices have been developed and calibrated against the observations. These models generally take into account the major meteorological parameters which govern the occurrence or non-occurrence of frost. These parameters are assigned weight according to their importance; means and standard deviations of these variables are generally used for the prediction of low temperatures. The indices based models present better forecasts than the direct maximum & minimum temperature-based models. In meteorology, there are a number of models which predict frost and hard freeze with good accuracy. But, weather services to which people have considerable access are generally quite broad and generalized as they use synoptic and/or mesoscale models to provide regional forecasts. Local or micro-scale predictions are typically unavailable. Therefore, imperial predictions which can be calibrated for local conditions are considered imperative. Further, most of the predictions tell about the minimum temperature which will be observed during the night. These predictions usually don't tell when the stress will start, how long will it prevails and what are the crops going to suffer. To address these issues a simple model depicting temperature evolution, duration of threat and the species which could be damaged, has been developed by Sharma [127] for the subtropical low hill and valley region of Himachal Pradesh, India. Rozante *et al.* [148] developed 'Frost Index' for pre-assessment of the frost threat at the regional level. He used the regional weather forecast model for developing the frost index. From his calibration process, he inferred that temperature has the largest contribution, followed by pressure, winds and other variables in influencing the frost and associated damage.

Prediction of temperature evolution and minimum temperature during a frost event

To predict the minimum temperature during a frost event and the trend of temperature evolution during a radiation frost event, Sharma[127] carried out studies with a number of meteorological variables like daily maximum temperature(T_{max}), temperature at the time of sunset (T_{530}), temperature drop within one hour after sunset (TD_{5630}), temperature drop within two hours after sunset (TD_{5730}), daytime minimum relative humidity (RH_{min}), relative humidity at sunset (RH_{530}), increase in relative humidity within one hour after sunset (RHd_{5630}) and increase in relative humidity within two hours after sunset (RHd_{5730}) etc. Different regression models were developed on the basis of these variables and were tested for their forecasting performance.

The descriptive statistics of the above said variables and their magnitude observed in the lower Himalayan region of Himachal Pradesh is presented here in the Table 4 for illustrative purpose. The meteorological averages (minimum or maximum or the daily mean air temperature or their confidence limits) presented in the table hardly give any indication of frost threat. But, in actual these are the limits of temperature when heavy damage due to frost was recorded in the region. Likewise, the straight information about maximum temperature averages or temperature drop rates or variation in humidity variables also fails to give estimates of potential frost damage threat. In order to develop appropriate prediction models, these variables were used as explanatory variables for predicting the progression of temperature toward minimum (T_{dip}). Several equations/ lines were developed using these variables. Some of the equations which were having highest value for the coefficient of determination, highest number of significant explanatory variables and significant F-Test are presented in Table 5.

Table 4: Descriptive statistics of the meteorological variables during radiation frost events in the lower Himalayan region

Variable	Averages			Confidence limit
	Min	Max	Daily Mean	
$T_{min\ (^oC)}$	-0.80	3.20	1.03	1.00 -2.33
$T_{max\ (^oC)}$	8.90	17.8	13.9	13.5-15.3
$T_{530\ (^oC)}$	7.10	12.7	10.9	9.5-10.3
$TD_{5630\ (^oC)}$	0.40	3.30	2.07	1.89-2.25
$TD_{5730\ (^oC)}$	0.60	5.90	3.48	3.18-3.78
$RH_{min\ (\%)}$	35.5	65.2	50.0	48.1-52.9

$RH_{530\,(\%)}$	46.0	72.3	67.4	43.1-48.9
$RHd_{5630\,(\%)}$	10.00	26.0	19.22	12.00-22.45
$RHd_{5730\,(\%)}$	16.00	36.0	24.7	20.3-32.9

Source: Sharma [127]

It can be inferred from that data presented in this table that regression line -I which was based on the sunset time (5:30PM) temperature (T_{530}) alone, explained 51.6% of the total variation in minimum temperature (T_{dip}) during a frost event. When temperature drop after two hours of sunset (TD_{5730}) was taken into consideration, the coefficient of determination became 0.6141; (Adj.R^2), means it explained 61.4% of the total variation observed in the minimum temperature. Further, adding the humidity variable *i.e.* increase in humidity two hours after sunset (RHd_{5730}), the forecasting ability of the equation further improved to 73.7%.

Table 5: Regression Lines for prediction of minimum temperature during a radiation frost event

Regress-ion Line No.	Regression lines	R^2	Adj. R^2	$F_{cal\,(0.05)}$
I.	T_{dip}=-2.06+0.285(T_{530})*	0.5556	0.5155	*
II.	T_{dip}=-1.86+0.426(T_{530})*-0.499(TD_{5730})*	0.6593	0.6141	*
III.	T_{dip}=-1.63+0.368(T_{530})*-0.558(TD_{5730})*+0.041(RHd_{5730})*	0.7518	0.7374	*

*significant at 5% level of significance

Though these linear regression models were statistically appropriate but, keeping in view the practical difficulty in taking observations on these variables at the farmers' level, it was decided to further simplify these equations into models of higher practical significance. Several polynomial expressions were tried considering different variables. Amongst the different expressions developed, one which is presented in the box below was found to have greater practical utility.

$T_i = T_{530} - (0.27\, T_{530}\, .i)/e^{(0.1i)}$ ...(Adj. R^2 = 0.6745)*... Reg. Line (IV)

$\int_0^{14} i$ = hours after sunset...{At sunset/5:30PM *i*=0, At Sun Rise/ 7:30AM *i*=14}

T_i = Temperature at i^{th} hour after sunset

$\int_7^{13} T_{530}$ = Temperature at sunset (5:30 PM)

*significant at 5% level of significance

This mathematical model expressed the temperature evolution during a radiation frost event right from the time of sunset upto the next morning under the frosty situations of the subtropical low hill and valley region of Himachal Pradesh. This mathematical model has advantage that it is based on only a single input variable i.e. sunset-time temperature and explained 67.5% of the variation in the dependent variable i.e. Temperature at the ith hour after sunset. This model was found capable of predicting the temperature evolution with respect to time, temperature drop rate and minimum temperature during a radiation frost event.

In a similar type of study Krasovitski *et al.* [149] while developing a 'Gaussian Identification Procedure', emphasized that the temperature drop rate after sunset definitely has a relation with the minimum temperature of a frost event. Also, Snyder and Melo-Abreu [8] illustrated updating of night temperature by taking pre-night temperature as a reference while predicting air temperature trend during a frost event. Their calculations projected the trend of air temperature as a square root function of temperatures from two hours after sunset until sunrise the next morning.

Evaluation of the forecasting performance of prediction models

Forecasting performance of regression lines III and IV (described above, illustration from Sharma [127]) having the highest value for the coefficient of determination (adjusted R^2) and the highest proportion of significant explanatory variables for prediction of minimum temperature has been illustrated in Figure 7 and 8, respectively.

Figure 7 depicts that the points of predicted versus actual values of minimum temperature are well distributed around the line of the perfect forecast. The calculated value of χ^2 also signifies the acceptance of regression line III for predicting the minimum temperature of a radiation frost event. Further, the 'Theil's Inequality Coefficient (U=0.317) also indicates that regression line III depicted very good forecasting performance for judgment of its forecasting ability as a model [8,150,151]. The partial coefficients of inequality indicate that the major contribution to the inequality,

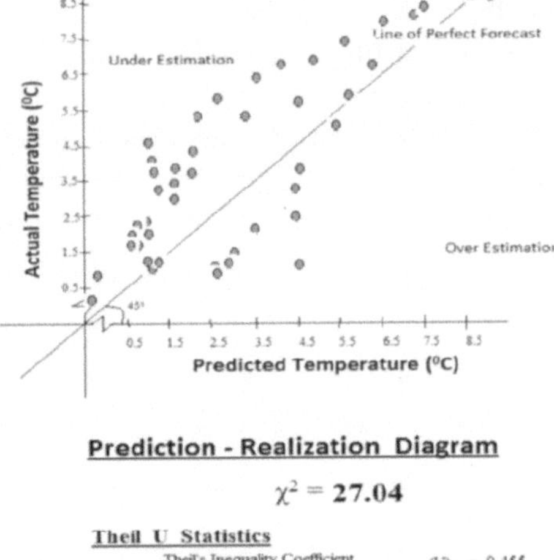

Fig. 7: Evaluation of prediction performance of estimated regression line

which existed during forecasting performance evaluation, was due to imperfect covariance (Uc =0.908). The systematic error (Um and Us) have contributed very little to the inequality which means that there are very less chances of improvement in the forecasting ability of this equation. In Figure 8, the Prediction - Realization diagram of regression line IV is presented. It is evident from the figure that points for predicted versus actual values were well oriented around the line of the perfect forecast. The majority of the points were in the 'underestimation zone', but the value of $\chi 2$ indicated acceptance of this equation as a prediction model for temperature evolution during a radiation frost event. Theil – U value also indicated the good forecasting ability of this exponential curve/line. The partial coefficients of inequality indicated that inequality existed mainly due to imperfect covariance (Uc) and unequal central tendency (bias proportion, Um). As per Theil [150,151], if the total inequality coefficient (U) cannot attain the desirable level 0 (zero), the most desirable level of bias proportion (Um) is near 0 (zero), the same is expected for the inequality due to imperfect covariance (Uc). As the predictions are rarely equal to actual outcomes, this type of error is of non-systematic type. Therefore, if Us≠0, this might be called 'systematic error' due to the forecaster's neglect. The distribution of inequality of line III is very near to desirable distribution i.e. Um≈ Us≈ 0 and Uc≈1, thus it has a greater level of acceptance as a model for prediction of low temperature during a frost event. However, regression line IV also possessed good forecasting ability but the value of Uc is a bit farther from 1(one) which indicates that the forecasting performance of this equation can be improved further in future with the incorporation of more information in the forecasting process. But, as this equation is based on a single explanatory variable and the data collection on this variable is easy and the forecasting ability of the equation is also good, therefore this equation was regarded as the best fit of practical significance for defining the temperature evolution during a radiation frost event under the N-W lower Himalayan subtropics.

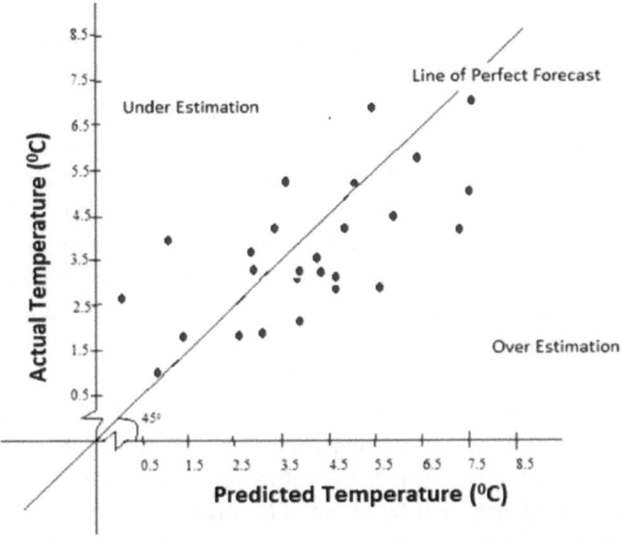

Prediction - Realization Diagram

$$\chi^2 = 11.84$$

Theil U Statistics

Theil's Inequality Coefficient	(U) =	0.317
Decomposition		
Proportion due to bias	(Um) =	0.80
Proportion due to variance	(Us) =	0.12
Proportion due to co-variance	(Uc) =	0.908

Fig. 8: Evaluation of prediction performance of estimated regression line 4

Frost protection guide chart

Most of the decisions pertaining to frost predictions and adoption of potential frost protection methods are done largely at the farm level by the farmers. In actuality, it has been observed that farmers' are generally unable to understand and execute the complex forecasting models and go for adoption of frost protection measures as per their assumptions. Due to this, the plantations either suffer heavy damage due to inadequate protection provided to their plantations or they apply heavy frost protection measures which either add to the cost of orchard management or impair plant health. It was therefore decided to further simplify the process of frost prediction and to make an assessment of potential frost threat in advance. The efforts were made to develop a simple prediction procedure so that farmer could foresee whether the coming night is safe for their plantation or not. And if, frost stress comes how long it will stay and what will be the critical time of the start of the stress. In order to address these concerns, a potential threat predicting chart called 'S- Chart' was developed in such a format that, a farmer will be required to know only the sunset (5:30 PM) time temperature and he can easily access the level of threat for his fruit crop (The S-Chart considers a frosty situation as 'threat' when 20% damage to the foliage of a plant can occur due to prevailing frost-induced freezing). The 'S-Chart' is presented in Figure 9 and it is very easy to read. There are presented three distinct coloured portions in the chart: Blue portion (I) represents the probable suffering of highly frost-sensitive crops like Banana, papaya etc.; Orange coloured portion (II) indicates the probable suffering of medium sensitive crops like jackfruit, mango, aonla, litchi, guava etc. together with highly sensitive fruit crops. The red region (III) indicates the probable suffering of almost all the subtropical fruit crops. From the chart, it can also be observed that if the sunset-time temperature remains above 11.7^{0}C, almost none of the subtropical crop will be affected by frost injury. At 11.7^{0}C the frost threat will start for sensitive crops like papaya and banana. At sunset temperature 9.3^{0}C the medium sensitive crops like mango, litchi, aonla etc. will also be under the frost damage threat together with the sensitive crops like papaya and banana. And if, the sunset-time temperature happens to be 8^{0}C or below, the frost threat will be there for almost all types of subtropical fruit plant species. Under the low hill and valley region of Himachal Pradesh, it has been observed that radiation frost threat happens only when the sunset-time temperature ranges between that 7^{0}C to 11.7^{0}C, at other temperatures, there have rarely been observed any frost threat in this zone.

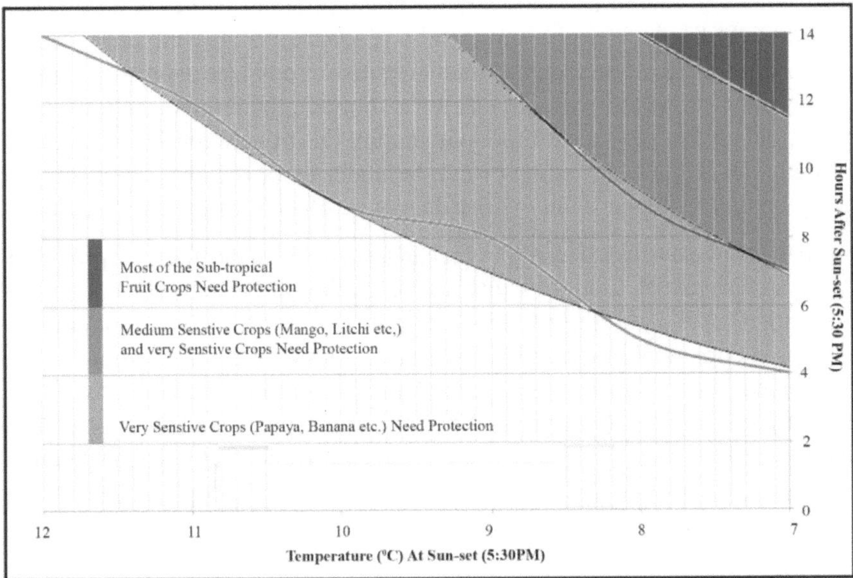

Fig. 9: 'S' - Frost Protection Guide Chart for Sub-Tropical Fruit Crops

This chart gives predictions very much like that of astrological '*Panchang*' which is being used by the pandits/jyotishies for making predictions of the '*Dasha*' (position/duration of influence) of the *'Grahas'* (planets). From S-Chart we can predict, at what time during the night the frost threat for a specific type of plants will start and what will be the duration of the threat. This chart, therefore, helps the farmers in deciding the level of protection they might need to fight the frost threat. For example, if the sunset-time temperature is 9.0°C, the S-chart tells that there will be a frost threat for sensitive as well as medium sensitive fruit crops. From the right-hand axis (number of hours past sunset), one can infer that for sensitive crops the threat will at around 6.5 hours after sunset i.e. at 12:00 AM and will last up to 14 hours after sunset i.e. up to 7:30 AM, in the morning. This means, the time after 12:00 AM is going to be critical hence the farmer may program his sprinkler system to operate after 12:00 AM. This will save water as well as the plants from root suffocation and will add to the operational cost of the orchard beside the wastage of water. Therefore, this chart is very user friendly. For medium sensitive crops, the frost threat will start at about 13 hours after sunset means the period of stress for such plants will of one hour only. Therefore, running of frost protection equipment is needed for one hour only.

For the clarity of the readers, one more example is elaborated herewith. Let the sunset-time temperature is 10°C. Here, the threat will be there for very sensitive crops only and this threat will start 8.5 hours after sunset i.e. at 2:00 AM in the midnight for the sensitive crops only. If a farmer is growing only medium sensitive crops and not growing frost-sensitive crops like papaya or banana, he can enjoy sound sleep during that night. Further, the farmers are required to bother only when the sunset-time temperature lies in between 11.7°C and 7°C, in the lower Himalayan region or maybe under other regions. The damaging frosty condition happens only under this range of sunset-time temperature.

From the above discussed examples, it can be concluded that the frost protection guide S-chart can be used for knowing the level of threat and protection needs of a target crop under the situations of potential frost threat. Automation of frost protection equipment can be facilitated with the help of this guide chart. It is a clear reality that in the years to come, the climatic variability will play even a greater role than in the past and therefore, precise predictions of frost and low-temperature stress will be of great significance in the years to come.

9

Delineation of Frost Prone Areas

As discussed in the previous chapters, the global perspective of frost is a latitudinal phenomenon; it is more common in subtropical latitudes (23.5^0 to 35^0 N&S). The proneness of vegetation to frost usually increases as we move toward the poles. But, at the micro-level, there are several factors like proximity to a water body, type of soil, type of vegetation, prevalence of winds etc., which govern the temperature inversion and frost-proneness of a situation or site. Above all, at the regional level frost is more an altitudinal factor and varies considerably with the altitudinal variation in an agro-ecological condition. Every watershed catchment experiences frost of different intensity under subtropical conditions; therefore, delineation of frost-proneness of the sites on the basis of revenue or administrative boundaries is not justifiable. Within the same watershed, a crop may suffer more frost damage at down-location sites than at a site having higher elevation. At the latitudes between the tropics of Cancer and Capricorn, there are comparatively massive areas with no sub-zero temperatures [152]. Even these tropical areas may experience frost injury typically at high altitude orchard sites. The length and intensity of frosty conditions at a specific site are the major criteria used in-vivo for defining the proneness of a site to frost. The geographical characterization of frost-prone sites at a regional scale is therefore difficult. There are only a few studies on this aspect of a frost risk delineation. A specific review on the geographical characterization of frost risk in Israel was published by Kalma et al. [153]. Based on the 25- year data and topographical information they developed maps showing a close relationship between elevation and the risk of sub-zero temperature under a subtropical environment. Other researchers used mobile temperature surveys or topographical and soil information, without temperature data, to derive risk maps. Although these maps are quite helpful to gather information on frost-free days, from a broader perspective, better spatial information on the risks of frost damage is still lacking.

There is no substitute for good local information and monitoring for accessing the vulnerability of sites for frost-proneness. Though, to change weather variables of a site is not possible but, systematic monitoring can help in delineation of the frost-prone sites and preventing the growers from considerable recurring losses from frost damage. In the N-W lower Himalayan region lying in between 300 to 1100 m above sea level, there exists considerable variation in the topographical features. Small basin areas are surrounded by converging and diverging ranges resulting in considerable altitudinal agro-meteorological variation within a single micro watershed. Thus high variability contributes considerably to the variation in frost intensity with respect to site elevation and other agro-ecological factors. To access the frost proneness of different agro-ecological situations, a delineation study was conducted by Sharma et al. [155]. The altitudinal variation, aspect, vegetative and other agro-ecological factors were taken into consideration while selecting different sites for frost proneness delineation studies in the lower Himalayan region of Himachal Pradesh. Continuous monitoring of the meteorological variables was done during the winters at the selected sites. The observations were also recorded on the natural frost damage to the frost-sensitive (papaya and banana), medium frost-sensitive (mango and litchi) and least frost-sensitive (citrus and pomegranate) fruit plant species [154, 155]. The quantification of the per cent frost damage to the fruit crop plants at a particular site was done by averaging the per cent frost damage observed to these three types of plant species at that particular site. Finally, the 'Frost Damage Score' for a site was arrived at by adopting the following formula:

Frost Damage Score (FDS) = [{(M+D)/2}/10]

Where,
M = % mortality observed in young plantations due to frost
D = % damage observed to the bearing orchards due to frost

Based on the critical difference observed in the frost damage score (FDS) and the local agro-ecological details, Sharma et al. [155] analysed the frost proneness of different sites under study. It was observed that all the low lying basin areas of the lower Shiwalik region experience both radiation and pool frost conditions. Sensitive crops like banana and papaya were found to suffer very heavy frost induced freezing damage. The medium frost-sensitive crops like mango, litchi, aonla etc. have also been found to suffer heavy to medium damage. Overall, almost all the subtropical crops were found to suffer frost damage to a variable extent depending upon their frost tolerance. The FDS value for these low lying basin areas varied from 6.2 to 8.6. Further, the plantation sites just above these basin areas also suffered heavy frost damage however, the damage observed was a bit lower than the flat basin areas. Both types of frost-sensitive and medium frost-sensitive fruit crops suffered frost damage in these areas also but, the damage to the medium sensitive crops was lower than that observed for the sensitive crops. The observations were found to be almost similar for the sites lying above or below 700 m above mean sea level (average elevation of subtropical low hill and valley region of Himachal Pradesh). The plantations on the Northern slopes also suffered a similar extent of the damage. The FDS value for such sites varied between 5.45 and 7.77. Moving upward along the slopes of the hills or hillock, it was observed that the extent of damage to different fruit crops was lesser than that was observed at the above discussed low lying sites. In the areas nearing the middle of the slope, only sensitive crops like papaya and banana were found to suffer damage due to radiative cooling. The incidences of convective or pool frost were rare at these elevations. In the case of medium sensitive crops like mango and litchi damage was observed only on the upper canopy of the trees. The economic losses to such fruit crops were low in these areas because of the draining down of the cool air, without giving any pooling effect near the lower or middle canopy of the trees. The FDS values of these areas were observed to be between 2.3 to 5.6 in most of the cases. The plantations on the upper portion of the hill or hillock were though found to experience only a little frost damage but the consistency was found to be variable from year to year and place to place. Due to the higher number of days with clear sky, these sites were comparatively found to be warmer during daytime and at the time of sunset in comparison to sites present at the lower portion of the slope or the basin areas. But, here temperature drop after sunset was found to be higher and the influence of cold wind blowing from the snow-bound mid-hill region was found to cause advection frost at these places. However, the frost damage to different fruit crops was found to be very low at these places. The FDS values for these areas were observed between1.4 to 3.7.

Further, the altitudinal variation in frost intensity within the sites selected for the studies was also monitored by taking observations at every 50 m of the vertical height along the slope in the areas both below and above 700 m of altitude. The observations on altitudinal variation in frost intensity are presented in Table 6. From the data presented in this table, it is evident that the sites which were above 700 m of altitude were comparatively less frost-prone than the sites which were below this altitude. Further, within a micro-watershed, heavy to very heavy frost damage was recorded in the plantations lying up to 50 m of vertical elevation from the lowest point of the micro-watershed. As we move up the extent of damage decreases with the increasing elevation under both below and above 700 m of altitude. The extent of frost damage was observed to be heavy only up to 100 m vertically from the lowest point of the watershed, at 100 to 150 m of elevation the extent of frost damage was observed to be medium. Above 150m, it was low to no frost situation at most of the sites studied.

Table 6. Altitudinal variation in radiation frost intensity in the low hill and valley region

Altitude(m) at the lowest point of the site	Elevation (m) of the site *w.r.t.* lowest point of watershed	Frost Damage Score	Frost Intensity
<700m	0-50 50-100 100-150 >150	>7.5 5.5-7.5 3.5-5.5 <3.5	Very Heavy Heavy to Medium Medium to Low Low to No Frost
>700m	0-50 50-100 100-150 >150	5.5-7.5 3.5-5.5 <3.5 <3.5	Heavy Medium Medium to Low Low to No Frost

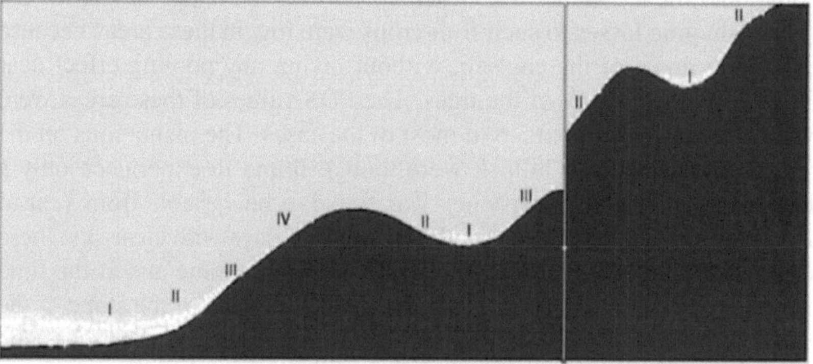

Fig. 10: Pictorial representation of agro-ecological situations and their frost pronness in subtropical Himachal pradesh

From the description given above, it is evident that the plain or basin areas at the basement of the hill or hillocks are very badly affected by frost. In comparison to the Southern slopes, the extent of variability in frost damage was observed to be lesser on the Northern slopes but the intensity was greater. On the Southern slopes, the extent of frost damage decreases as we move up along the slope. Based on above discussed observations, the frost proneness of the low hill and valley region of Himachal Pradesh has been depicted pictorially in Figure 10, delineating this whole zone into four distinct Agro-ecological Situations (AES). The frost proneness description of these four AES is presented below:

AES-I: It is highly frost-sensitive and depicted here in the figure at the basement of the hills. The vertical range of this AES is up to 80 to 100m from the lowest point of the watershed.

AES-II: The frost sensitivity of this AES is high. This AES lies above 100m from the lowest point of the watershed. The vertical elevation of the sites may be more than 100m on the Northern aspects.

AES-III: The sloppy area nearing the ½ of the sloppy terrain has been put under this AES. The frost sensitivity of this AES has been ranked to be medium.

AES-IV: The upper portion of the hill or hillocks lying below 1000m above mean sea level has been categorized into this AES. These areas comparatively experience few incidences of radiation and advection frost. The cool air usually drains down from this area so temperature inversion help in the prevention of severe frost damage to the fruit plantation.

In the nutshell, it can be concluded that the frost vulnerability of a site in the low hill and valley regions is related more to the altitudinal position of the site within the watershed. The higher we move within a micro-watershed/ watershed, the lesser is the frost-proneness of a site. Therefore, while planning commercial plantations in the topographically variable sites, it is always better to consult the local growers regarding the altitudinal variation of frost intensity within the watershed; they usually possess good knowledge about the low-lying spots where cool air pools or where the advection frost is a problem. As the ground fog form first in the low-lying basin areas, a good rule of thumb is to avoid places where ground fog forms early in the winters. Further, one should review local topographical maps before planting frost-sensitive crops on high-risk sites.

10

Avoidance and Edurance of Frost

The problem of frost damage to the crops is as old as agriculture is. Earlier, crop production was undertaken only as per the geographical distribution of the plant species and the cultivation of crops in non-conventional areas was not there. Later, with the expansion of agricultural activities, a number of introductions were made in the non-conventional areas. But, the survival of the introduced crops remained at the mercy of Nature God. These introductions faced a number of challenges; the frost induced freeze damage is the major one that restricted the geographic distribution of the plant species across the globe. Due to heavy frosty conditions, citrus cultivation has been reduced considerably in South Carolina, Georgia, Northern Florida regions of the Uniited States and other parts of the World [156]. The mango cultivation in the frost-prone subtropical NW India has been considerably hampered during past twenty years and therefore, the subtropical region in the NW Himalayas lost its priority for mango cultivation due to frost [154-155].

Earlier, frost was believed to be showered from the sky like snow during the clear nights and the farmers used to cover their plants or crops with thatch for mitigating its devastating effects. Later, the role of these thatch covers was realized as the thermal insulators used to reduce heat loss and give protection to the plants against freeze damage. Thatch covers are still common in almost all frost hit areas of the world despite their limited capacity to provide protection.

Systematic studies on exploring frost and its impact mitigation started as early as 1932 when Dexter *et al.* [131] described a methodology for quantification of frost damage through cellular ion leakage studies. They used this method for illustrating the relative hardiness of different plant species to frost. Regarding the geographic variation in frosty conditions, Bagdonas *et al.* [157] presented a detailed account of the impact of frost under tropical, subtropical and temperate regions.

During the nineteenth century use of heaters began which used the burning of heavy oils or old rubber tyres together with sawdust to create smog to prevent radiation losses from the orchard. During the middle of the 20th century, the use of wind machines started and replaced the costly and environmentally unsafe heaters. Though the wind machines were costly to install, the labour and operational costs were lower. Later in the eighties, new technology of using aqueous foams against radiative night cooling was developed [149,158,159]. In this technique; orchard heat transfer is restricted by spreading a foam layer over the orchard. But, due to its bulkiness and poor cost-effectiveness this technique could not get popular. In recent times, the use of methods such as fogging and overhead sprinkling has been started but the efficacy of these methods is still questionable.

As we know frost is a serious natural threat; mere adoption of a protection method rarely gives sufficient protection against its devastating effects. Therefore, knowledge about frost occurrence and its mechanism of damage under varied agro-climatic situations is essentially required for mitigating its ill effects on the subtropical fruit crops. The methods used for protection against frost are categorized into passive and active methods depending upon the type of frost protection they provide. The practices which help the plants to avoid and endure frost stress are called **Passive Frost Protection Methods**. Whereas, the method which provide direct protection or act as a shield against frost threat are known as **Active Frost Protection Methods**.

Some of the important frost protection methods which are used for avoidance and endurance of frost under subtropical conditions are discussed here:

1. Avoidance of Frost

To avoid or escape the threat is the best way to deal with it. For this, the orchardists should gather sufficient information on frost vulnerability of the plantation sites and crop species right at the time of planning of an orchard. The pre-planting frost avoidance strategies are comparatively less costly than the frost endurance methods. If practised wisely, these strategies often eliminate the need for the adoption of any post-planting frost protection method. Some options are discussed here under which growers can follow for avoiding losses caused due to frost in the subtropical regions.

i) **Selection of the frost-free site:** Proper site selection is one of the most important ways for avoiding frost damage. The majority of the losses occur at sites that are not selected properly and frost vulnerability of which is ignored. It is a fundamental principle that the denser cool air flows downhill and pools in the low-lying pockets. These pockets initially experience radiative cooling. With the enhancement in the number of nights with freezing temperature, the accumulation of cold air (convective cooling) keeps these sites drenched with cool air leading to devastating effects. As discussed in the previous chapter, the intensity of frost decreases as we move up along the slope in hilly areas. Local growers and extension workers often possess good knowledge regarding the frost-proneness of the sites in the locality, therefore, their opinion can be taken into consideration while gathering information on the frost vulnerability of the site before making the decision final. Supportive information on a delineation of frost-prone area and prioritization of crops for these areas can be had from Chapter-9 for better decision making.

Frost sensitivity of the sites usually varies due to variation in soil type. Different soils possess a different capacity in respect of conduction and storage of heat, therefore, add variable levels of frost vulnerability to different sites. Soils of coarse-grained texture such as gravels and sands possess high conductivity therefore, the dry sandy soils transfer heat in a better way than the dry heavy clay soils. The heat is therefore, transferred and stored better in this type of soils. Organic soils on the other hand have poor capacity to transfer and store heat regardless of the water content hence more prone to frost damage. Further, the depth and rate of frost penetration into the soil and further rate of thawing of frozen soil also depends upon the thermal properties of the soil. The difference in conductivity of frozen and unfrozen soil chiefly depends upon the moisture content. The sites with different soil water regime behave differently as far as the conduction and storage of heat is concerned. When the water content of the soil is near field capacity, soils reach a condition that is most favourable for heat transfer and storage. Under shorter spells of freezing, wet conditions of the soil are quite preventive in frost damage. Under the conditions of prolonged freezing air temperature persistence, the whole amount of water in the soil freezes spontaneously just after the occurrence of ice nucleation. Most plants under such situations die of desiccation due to the freezing of available soil moisture. Therefore, under frost-prone situations apart from the soil type and soil moisture regime, the prevailing frequency and duration of frost spells also influence the frost vulnerability of the sites.

The aspect of the slope also has a role in the frost vulnerability of the site. The Northern aspects are always more vulnerable to frost than the similar sites on the Southern aspect. If multiple crops are to be planted, it is best to plant deciduous crops on North facing slopes of hills and to have delayed springtime bloom. The probability of freezing decreases rapidly with the passage of time in the spring. The deciduous crops on south-facing slopes bloom earlier. As a result, deciduous crops on south-facing slopes are more prone to freeze damage. Subtropical trees (e.g. mango and litchi) are damaged by freezing regardless of the phenological state, so they should be planted on south-facing slopes where the soil and crop can receive and store more direct energy from sunlight.

ii) **Selection of frost-tolerant germplasm:** Another major method for avoiding the devastating effects of frost is the selection and planting of frost-tolerant crop germplasm. Establishing robust and reliable frost tolerant phenotypes always prove helpful for avoiding big production losses in subtropical horticulture. Unfortunately, very little has been known about the frost tolerance of subtropical fruit crops and their varieties. The limited advances in the development of frost-tolerant germplasm are due to the practical difficulties in measuring frost damage under field conditions due to its sporadic and dynamic nature. The availability of frost-tolerant parental lines is also limited which can be used for resistance breeding.

During the past few years, considerable information has been generated on the relative frost susceptibility of some of the subtropical crops which is discussed in detail in Chapter-6. The relative order of frost susceptibility of different crops is presented in this chapter and the information on prioritization of fruit crops for the frost-prone subtropics can be customized as per requirement at the time of orchard planning.

Frost not only affects the vegetative growth of the plants, flowering of some of the early blooming and/or terminal bearing crops also gets affected. For the frost-prone sites, it is advised to select crops and varieties which bloom late. For example, low chilling peaches and grapevines grown in the subtropics do not suffer frost damage to the trunk, branches or dormant buds, but they do experience damage as the flowers and small fruits develop during a frosty situation. Selecting deciduous plants that have a late-blooming habit provides good protection because the probability and risk of frost damage decrease rapidly with the onset of spring. Most citrus species tolerate temperature up to -2^{0}C, if it persists for a shorter duration. Mandarins, Tangerines or Satsumas are cold hardy than the common sweet oranges. Mandarins cannot be grown in the mountains or areas with continental winters, but with precautions against sudden temperature drops below freezing, they can typically be grown. The most cold-hardy mandarins are Satsumas and Changsa which can survive temperature as low as -9^{0}C [160, 161]. Plants of seedling origin are more cold-hardy than the grafted ones. In Texas, USA, 'Satsuma Orange Frost' is the name given to a well-performing highly frost tolerant Satsuma mandarin variety of the region [161]. In comparison to tangerines and mandarins, sweet oranges and grapefruit are less cold hardy. The oranges tolerate freezing temperature stress to some extent but excessive negative temperatures are quite detrimental for sweet oranges. These can be grown in a cool climate up to -7^{0}C if adequate protection is provided [162]. Apart from selecting a frost tolerant scion variety; rootstock selection is also very crucial in citrus. Trifoliate orange rootstock is considered a better choice for citrus growing under frost-prone situations.

In the case of terminal bearing genotypes, considerable damage occurs to the reproductive buds due to frost and thereby flowering and fruiting gets hampered. Therefore, in the case of terminal bearing types, canopy morphology play important role in frost damage avoidance. It has been observed that the entire top canopy usually shields the underneath canopy from frost induced freeze damage under radiation frost conditions. It has therefore been recommended that for planting under radiative cooling situations, preference should be given to the strains/ varieties of fruit crops that have an entire and denser upper canopy. Further, pruning of the upper layer of the canopy should be avoided under such frosty situations as it may aggravate the frost damage upon removal of shielding effect giving canopy area.

2. Endurance strategies for frost impact mitigation

Fruit orchard is a fixed enterprise, once the orchard is planted it is not easy to change, it costs the grower both money and time. Therefore, every care must be taken in the selection of appropriate site and planting material for the frost-prone sites. If the planting in frost-prone sites is done without giving consideration to these factors, it may incur recurrent losses to the orchardists. Under such situations, the growers are left with no option but the endurance of frosty conditions. Discussed below are some of the post-planting frost endurance practices which can be adopted for mitigating the devastating effects of frost.

i) **Improvising cool air drainage:** Often, the most severe frost damage occurs when the micro-scale advection happens. A careful study of the topographical map of the site is always helpful in avoiding frost-prone situations. Illustrative information on the down slope flow pattern of the cold air can be generated by creating plenty of smoke during a frosty night, not necessarily when the temperature is subzero. Once the pattern of cold air drainage is known, the planting pattern can be decided accordingly. The freeze potential of the site is generally higher where there are certain restrictions like bushes, buildings, bales of hay, or dense vegetation etc. restrict the air drainage downhill.

The best way of reducing the freezing potential of these sites is to modify the course of the flow of cool air just from the outer sides of the orchard. Protective Shelter belts, bushes, stacks of hay, mounds of soil, buildings, boundary walls, berm walls, fences along the upper periphery of the orchard proves very helpful to funnel cold air around the orchard and escaping the frosty situations. Further, the orchard row orientation should be such that it facilitates the drainage of cold air. Under high slope conditions, this practice is not advised as it will increase soil erosion. Under such situations due to the high slope of the site, the cold air rarely gets restricted and therefore row orientation may be taken obliquely to the slope for keeping the soil erosion at its minimum. If *per se* adoption of these measures is not possible, the restrictions in the down flow of cool air should never be allowed for avoiding micro-scale advection or pool frost inside the orchard site. In the orchards where air drainage is problem, the use of air blowers can be made for churning the cool air and avoiding the frost induced freeze damage to the plantations.

ii) **Shelter belts:** Under the flat orchard site condition where advection frost is a problem, shelter belts are quite effective for restricting the mass flow of cool air into the orchard. Single or multi-row plantation of tall trees along the periphery of the orchard for the purpose of protection of orchard from fierce weather is called Shelter belt (wind-break). The shelter belts are also planted for the reduction of soil erosion, salinity control and biodiversity improvements. It moderates the temperature of the site and acts as a barrier for the movement of cold air mass from the surrounding environment to the orchard site. The impact of the shelter belt may be modified by its height, density, number of rows and species composition. Length, orientation and continuity of the shelter belt also affect the weather variables of its surroundings. Under frosty situations, the shelter belt imparts protection to the orchard by restricting the mass flow of the cold air from outside. The significant direct effect of shelter belt can be observed up to 1.25 times the height of the trees planted as shelter belt under non-windy and up to twenty times under windy situations. Apart from the height of the shelter belt, the continuity and thickness of the belt also matter for providing protection against frost. It has been observed that for maximum efficiency, the uninterrupted length of the shelter belt should be at least 10 times its height.

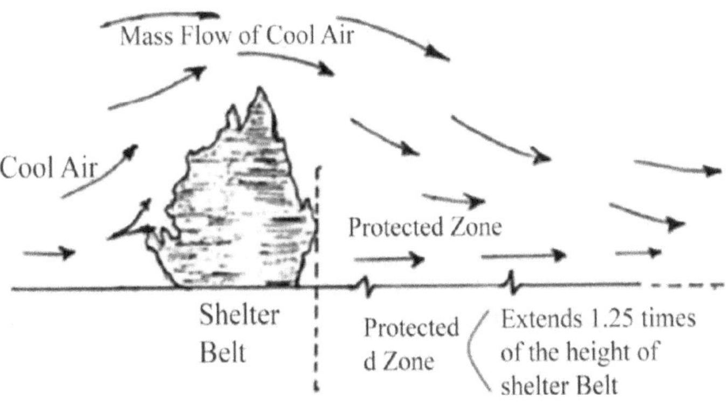

Fig. 11: Effect of Shelter Belt on Mass Flow of Cool Air and Frost Protection

iii) **Canopy tree plantations** are those plantations that comprise economically valued trees planted in between the orchard just to protect the underneath orchard plants against frost and low-temperature stress. Date-palm intercrops are common in Southern Alabama because the date-palms impart certain level of protection to the citrus orchards growing underneath[180]. Because, the dates also give a marketable product, this system of cropping is quite popular over there for achieving frost protection together with additional income generation. Under Indian conditions, in Punjab some of the growers interplant poplar trees in between the young Kinnow mandarin orchards. The poplar trees enhance the long-wave downward radiation and impart protection to the mandarins. Similarly, shade giving trees are also used to protect coffee plants from frost damage [8].

iv) **Removing trash from orchard surface:** The presence of plant residues inside the orchard also enhances the frost-proneness of the sites. As per Snyder and de Melo-Abreu [8], the air temperature measured within grape vineyards and citrus orchards with plant residue or grass cover typically vary between 0 °C and 0.5 °C colder than grape vineyards and citrus orchards with bare soil, depending on the soil and weather conditions.

Thus avoidance and endurance of frost are the important aspects of stress mitigation. Identification of the basic causes temperature inversion, cool air pooling and orchard energy losses can help in a great way for understanding the functional mechanisms of plant responses to frost and low temperature stress. Large chunk of orchard management cost and losses thus can be avoided by paying attentions to the above discussed frost stress mitigating practices at proper time.

11

Frost Tolerance Development

One of the important strategies to mitigate the adverse effects of frost is to develop tolerance in plant against frost stress. Tolerance is such a condition of frost vulnerable plant species, they develop certain ability to withstand the freezing stress by avoiding the states of inter or intracellular freezing. Thus, the process of cold hardening or acclimation together with ability to super cool at freezing temperatures is usually termed as frost tolerance development. Plants may be induced to tolerate freezing stress to a certain extent by:

- Avoiding freezing through depressing freezing point and increasing the degree of super-cooling [164],
- Restricting extracellular freezing by reducing the amount of ice formed due to an increase of the concentration of solutes in the protoplasm [165],
- Improving tolerance to a higher degree of desiccation due to the plasmolysis of the protoplasm[166],
- Increasing the permeability of the plasma membrane to avoid intracellular freezing [11],
- Reducing free water at plant surface for restricting ice nucleation and spread;
- Modifying cellular lipid-water proportion.

Thus, the development of cold tolerance is a result of highly complex processes including the induction of genes encoding stress proteins (e.g. dehydrins), increase in sugar concentration, enhancements of antioxidative mechanism and changes in lipid and protein composition which has some relation with the receding temperatures [167,168,169].

Cold hardening and acclimation

The term cold hardening is generally used in synonymy with cold acclimation which is more a genetic feature of a plant species. The natural process of regulating the physiological mechanisms so as to build tolerance to low-temperature stress in winters is called acclimation. It induces a set of biological processes that a plant undergoes to prepare itself for the cold weather. Biochemical and physiological changes in the intrinsic mechanism of the plants occur to avoid or mitigate cellular injuries caused by sub-zero temperature. Whereas, the plants of tropical origin like papaya, mango, banana etc. do not possess sufficient intrinsic mechanism for cold hardening and therefore find it difficult to survive freezing temperature. However, these species try to acclimate to some extent by way of regulating their water and other biochemical contents with the receding temperature. The subtropical plants which show certain degree of tolerance to low temperature stress develop some biochemical and physiological modifications in their system to withstand the gradually receding temperature. The gradual exposure of the plants to non-freezing low temperature is thus beneficial for the evergreen plant species to attain a certain level of acclimation to the low-temperature stress. The biomolecular mechanisms associated with the process of acclimation have been discussed in detail in Chapter 5. Under the changing climatic scenario sudden rises in temperature during winters are not uncommon. And, this abrupt change in temperature during winter is quite detrimental to the acclimation process as it de-acclimatizes the plants. This rise in temperature does not sustain longer and the temperature again recedes and keeps on receding. This reversal of low temperature causes heavy damage as the plants do not get sufficient time to re-acclimatize. The process of re-acclimation is usually slow in many plant species. The intermittent rise of temperature induces new growth which suffers more damage. The process of hardening is a bit different in temperate region plants; these plants adapt to winter and sub-zero temperatures by relocating the nutrients from leaves and shoots to storage organs. These plants shed their leaves and bring their physiological activity to a minimum so as to escape the low-temperature stress, this adaptive mechanism is missing in tropical and subtropical plants species.

Super-cooling: It is the process of cooling a plant or plant organ below freezing temperature without the formation of ice crystals in the inter- or intra-cellular spaces. The apoplastic water, solute content and cell wall rigidity plays a great role in super-cooling and freezing. Arias *et al.* [166] found that leaves of cold-acclimated olive cultivars can super-cool to -13^{0}C, substantially lower than the minimum air temperatures. The cold-acclimated cultivars with higher osmotically active solute content, higher tissue elastic adjustments and lower apoplastic water possess a higher super-cooling ability. The gradual exposure of the plants to receding temperatures also enable them to switch on their biochemical mechanism for increasing their solute concentrations and get super cooled without experiencing the cellular freezing damage. Hence, the process of super-cooling is quite helpful to the plants for building tolerance to freezing temperatures.

Regulation of lipid and water content: Freezing injury in plants is primarily a consequence of membrane destabilization resulting from freeze induced dehydration [171]. Tolerance to low-temperature stress requires synchronous operation of several plant processes during acclimation for increased stability of cellular membranes during freeze induced dehydration. The stability of the cellular membranes during freezing is further dependent on a number of functional and structural characteristics of these membranes. The fundamental structure of the plasma membrane is the phospholipid bilayer which forms a stable barrier between the aqueous compartments of the cell. Recent studies suggest that lipids are the fundamental structural unit of the cellular membranes and largely responsible for membrane integrity under stress conditions. They (lipids) have special significance in the freezing tolerance of the plants due to their direct association with membrane plasticity [42]. But, as stated above the freezing process is also associated with cellular dehydration, the injury due to freezing do have linkage with the hydrological status of the plant or plant organ suffering freezing. As a measure of acclimation, the plant species try to regulate their lipid and/or water status under low-temperature stress so as to improve their capacity to withstand stress, may it be of one or a few degree Celsius. The dynamics of water and lipids in plant tissues during winter and prior to winter was studied in some subtropical fruit species and it was found that most of the plants start regulating their water and lipid content prior to winters as they receive environmental cues of temperature recession [169]. The leaf water content was observed to recede gradually with start of autumn in almost all the species studied. This decrease progresses at a negative exponential rate with the progression of winters. Further, the leaf water content of deciduous species recedes faster than evergreen species which continue to decrease their leaf water content slowly upto mid of winters at a decreasing rate of water loss. By the end of the first week of November, the deciduous subtropical species like bil, lasura and phalsa reduce around ¼ of the total water loss from their leaves. Afterwards, only a marginal decrease occurs upto leaf abscission.

Among the evergreen subtropical fruit plants studied, it was found that a significant reduction in leaf water content of karonda, guava, galgal, aonla, raspberry and ber occurs up to the first week of December (or even up to mid of December depending upon the prevailing minimum temperature). On the other hand, in fruit plant species like custard apple, papaya, litchi, mango, banana, passion fruit, and jamun etc. reduction in leaf water content was not significant. As far as the frost tolerance of these fruit species was concerned, aforementioned species (which reduced their leaf water content) possessed better frost tolerance than the later ones which have not shown significant reduction in their leaf water content[123,155]. These studies revealed that the evergreen fruit species possess differential behaviour for modulation of leaf water content in relation to receding environmental temperature. Mostly, the species which could reduce their leaf water content possessed better tolerance to frost than those which did not show signs of significantly down-regulating their leaf water content. But, it was observed that there existed some species which do not regulate their leaf water content and still had better tolerance to frost.

Unlike the above described fruit plants, the species like lime, sweet oranges, mandarins, strawberry, pomegranate etc. were found not to down-regulate their leaf water content but they also did not suffer significant frost damage. The relationship between frost tolerance and leaf water content of these species could not be established. Earlier studies conducted on these species have also shown their better frost tolerance than the other evergreen fruit plant species[155]. It has been speculated that there might be present some other mechanisms with a stronger role in their frost tolerance[169]. Studies on the seasonal variation in the total lipid content of the leaves of these species provided additional insight into the phenomenon of their frost tolerance. The studies revealed that these species were efficient enough to up-regulate their leaf lipid content with the start of temperature recession. So it was concluded that seasonal variation in leaf water and lipid content of the fruit plants play a role in the frost tolerance of subtropical evergreen species. Thus, any strategy or method which can help in the regulation of leaf water and lipid content may prove of high worth for developing versatile frost protection mechanism and to avoid the freezing stress in subtropical fruit plant species.

Improving Plant Nutrition: The onset of short days and cold temperatures trigger the recession of growth and development of almost all plants. The farm practices which stimulate growth in the autumn encourage the emergence of late flushes and increase the vulnerability of the plant species to frost or low-temperature stress. As plant nutrition has a direct role in transmitting the growth stimuli and further maintaining the growth rates, the frost endurance potential of the plants is thus greatly influenced by their nutritional status. For example, nitrogen application late in September encourages a late flush of vegetative growth that plants can't turn off well before winters and therefore heavy frost damage is observed in such plants. Contrarily, the herbaceous perennials require good nutrient status right during the fall for stimulation of vigour and storage of more food for winter which often help these to survive better and re-grow more vigorously in the next spring. However, in the case of woody perennials weak and unhealthy plants are more susceptible to frost damage than healthy ones. Timely manure and fertilizer application are therefore required for these plant species for maintenance of plant health and to reduce their vulnerability to frost damage during winters. Also, plants that are poorly fertilized tend to cease their growth earlier in the autumn and bloom earlier in the spring; it further increases their vulnerability to frost damage. Therefore, balanced nutrition of the plants is of utmost importance for having reduced vulnerability of the fruit orchards to frost stress.

Literature on the relationship between specific nutrients and their role in low temperature or frost stress endurance is obscure; most of the information available either have partial interpretations or contains many contradictions. Still, there are some unbiased reports which suggest that in general, the nitrogen and phosphorus fertilization before winters encourages growth and increase the susceptibility of the plants to frost damage. Therefore, the fertilizer recommendations of most of the advocate that to enhance the hardening and frost endurance of plants, it is better to avoid applications of nitrogen fertilizer during late summer or early autumn. However, as phosphorus is important for cell division and helps in the recovery of the affected tissues after freezing, its application is not restricted. Likewise, potassium exerts favourable effects on water regulation, photosynthesis and stress tolerance in plants therefore, its pre-winter application is also not restricted. But, researchers are divided about the benefits of potassium for frost protection [8].

Slowly mobilized fertilizers like superphosphates, muriate of potash or manures and lime application (according to soil test results) are recommended for application in the subtropical fruit orchards in October or November. In winters, when temperatures are cooler and days are shorter, these fertilizers do not stimulate top growth until the spring as the availability of nutrients from these fertilizers takes some time. Even if these become available and the roots take up, these nutrients get stored in the roots and stems. Therefore, it is best to apply these fertilizers, manures and lime in the fall. If somehow the timely application is skipped, it is better to wait until growth is resumed in the ensuing spring. Calcium has been reported important in low-temperature stress tolerance as it triggers the production of secondary messengers for induction of the acclimation process in the plants. Calcium ions are perceived as secondary sensors in plants to cold stress responses [172]. These ions act as mediators of stimulus-response coupling in the regulation of plant growth, development, and responses to environmental stimuli [173,174]. Cold stress-induced rigidification of plasma membrane at microdomains causes cytoskeletal rearrangement which is followed by the activation of Ca^{2+} channels and increases cytosolic Ca^{2+} levels, which gets involved in the cold acclimation process [175,176]. Therefore, maintenance of optimum Ca^{2+} level in the cellular reserves is essential for the development of acclimation against low-temperature stress.

Iron and copper are considered important in managing oxidative stress under low-temperature conditions. It is generally assumed that HO· in biological systems is formed through redox cycling by Fenton reaction, where free iron (Fe^{2+}) reacts with hydrogen peroxide (H_2O_2) and catalyzes the Haber-Weiss reaction that results in the production of Fe^{2+} ions [177,178]. The hydroperoxyl radical (HO_2.) plays an important role in the chemistry of lipid peroxidation which is associated with the integrity of cellular membranes [179]. The role of iron is therefore important in cold stress management in plants. The role of copper is important in antioxidant response and accumulation of sugars. The role of other nutrient elements is not much clear in cold stress tolerance.

12

Active Methods of Frost Protection

When avoidance or tolerance of frost is not possible and frost damage is inevitable, the active methods of frost protection are the way out for the protection of the fruit plantations. The methods which are employed just before or during a frost event for avoiding freeze damage are called **Active Frost Protection Methods**. Under calm non-windy situations, the active methods of frost protection are very helpful especially under situations when the air temperature is above the critical damage temperature of the crop plants. Active protection methods are usually effective against radiation frost conditions. Under pool frost conditions also these methods give sufficient protection when pooling is very long in persistence. For attaining effective frost protection from the active protection methods, it is very essential to have good forecasts of on-farm minimum temperature and wind conditions. Knowledge of critical temperature of crop damage, night time temperature evolution trend, information on duration and severity of frost all are always helpful for attaining active frost protection. These methods though have certain limitations like, may cause more frost damage under windy conditions, the operational cost is high; still these methods provide immediate protection against freezing under most of the frosty situations. Some of the common in use methods of active frost protection are discussed hereunder:

1. Water Sprinklers

As the water cools, it loses thermal energy into the surroundings because the particles in its liquid state have more energy than in the solid state; therefore, energy is released as the temperature of the water drops to freezing point. This property of water has been used since ancient times for the protection of plants from frost damage. Twenty calories of energy are released into the surroundings when the temperature of 1g of water falls from 20^{0}C to 0^{0}C. Further, if it freezes at 0^{0}C it releases 80 calories in addition. Therefore, the sprinkling of water is widely used for protection against frost. There are many sprinkling methods that are used for frost protection, major ones are described here:

i. **Over-Head sprinklers:** This method has been found to be very effective in frost protection of evergreen subtropical plant species especially under situations where wind velocity remains below 8 km/h during night time. The major advantage of this method is that the energy consumption for operating a sprinkler system is considerably less than the energy used by heaters for attaining a similar level of frost protection. The labour requirement for operating a sprinkler system is also lesser than that used under other methods of active frost protection. Above all, it is relatively a non-polluting method and can be used under radiation and advection frost conditions up to -2^{0}C, if the application rate is sufficient and the coverage is uniform. However, under windy situations or prolonged persistence freezing temperature, this method brings adverse effect. Heavy frost damage to the plants may occur under windy conditions as the sprinkled water gets evaporated. The evaporative cooling worsens the situation. Under prolonged freezing situations, the amount of heat added by the sprinkled water becomes non-significant in comparison to the persistence of lower temperature and thus it gives little protection to the plants. Further, continous operation of sprinkler system for most number of days make the soils water-logged and suffocates the plant roots. Other drawbacks of using overhead sprinkling as a frost protection method are:

- Severe damage may occur to the plants if the sprinkler system fails
- A larger volume of water is required in this method
- Ice loading can cause branch damage if ice formation occurs
- Root diseases become a serious problem in poorly drained soils
- Variable application rates are required for different types of sprinklers to be used.

In addition, the frost protection through this method is achieved only up to the situation when the liquid-ice mixture keeps on dripping from the plant. If a lesser amount of water is applied, the freeze spread fast onto the plant surface. Therefore, for effective protection the over-plant sprinklers should cover all the plant parts with a thin water layer and re-wet these after every 30 to 60 seconds. Also, in the case of bigger trees for having larger coverage, the overhead sprinklers are required to be mounted at least 1m above the canopy top for getting broader coverage. For protecting the sprinkler heads from icing, specially designed springs are often required which add to the cost. Clean filters are needed to be sure about the proper operation of the sprinkler system during its need.

To reduce the time of sprinkler operation, the decision regarding when to start and stop the sprinklers for frost protection should be based on both temperature and humidity persisting in the orchard. Whenever a sprinkler system is to be started, it should be kept in mind that the orchard air temperature will first fall to the wet-bulb temperature. This initial drop is usually followed by an increase in temperature as the water cools further on the ground and plant parts. However, if the dew-point temperature is low, then the wet-bulb temperature can be considerably lower than the air temperature and the initial temperature drop may lead to damage. Therefore, starting and stopping of sprinklers for frost protection should always be done when the wet-bulb temperature is above the critical damage temperature of the crop. Automation of the sprinkling system should therefore be adjusted on the basis of the critical lethal temperature of the crop under field conditions and the expected wet-bulb temperature. Under most of the situations the sprinkler should be started when air the temperature reaches near 0°C. It should be operated till water starts dripping from the plant surface and should be restarted for 30 seconds after every one hour under the subtropical orchard conditions where minimum air temperature rarely fall below -2°C.

The extent of use of water during over-head sprinkling, with a conventional sprinkler, usually depends on the prevailing wind speed, minimum temperature and rotation rate of the sprinkler head. The water application rate is generally increased with the increase in wind speed. At lower atmospheric temperature higher application rate of water is required. Similarly, when rate of rotation of sprinkler head is slow, application rate of water is required to be raised. As a rule of thumb we consider that rate of water application is better which result in dripping of clear water or water falling down as a mixture of liquid-ice. The water application rate is considered low, if, the sprinkled water freezes at the foliage and gives milky-white appearance like rime ice. At these water application rates heavy frost damage may occur to the fruit orchards. Therefore, the application rate should be sufficient to adequately cover all of the foliage otherwise, damage can occur on plant parts that are not adequately wetted. Under windy situations where high evaporation conditions prevail, inadequate application rates can cause more damage than if the sprinklers are not used. The effectiveness of the sprinklers thus depends greatly on the evaporation rate, which increases usually with the increase in wind speed and with a drop in dew point temperatures.

The best way for automation of the sprinkler system is to follow the guidelines of the 'S-Forst Protection Guide Chart', which are discussed in detail in Chapter-8. The automation of the system may be tuned to the time when there is a potential frost threat to the orchard. When a frost threat is not there for a particular plant species, it is better not to operate the sprinkler system to prevent water logging of the orchard soil.

ii. **Under-plant micro-sprinklers:** This method of frost protection is also popular among growers under the mild to medium frosty conditions. More protection is achieved through this method due to lesser chances of water evaporation than the overhead sprinkler method. It operates on low pressure therefore, it is more affordable than an overhead sprinkler system. This method uses lesser water but, the best protection is achieved when sufficient water is used to cover a larger area. One to two spray-jet micro-sprinklers are generally installed per tree at the ground level or on short risers. Unlike overhead sprinklers, micro-sprinklers do not commonly wet leaves and branches above a height of about 1m and do not usually cause serious limb damage. For the purpose of frost protection, a micro-sprinkler system is needed to be designed for providing water to the entire orchard block all at once. If designed for a smaller proportion of the orchard, its effectiveness is reduced. The under-canopy sprinklers are usually less effective when temperature drop rate is high under radiation frost conditions.

iii. **Early Morning Water Sprays:** During a radiation frost event, the minimum temperature usually occurs during the early morning just on hour before sun-rise. After that the temperature starts rising sharply with the progression of time. Under the conditions of sharp temperature rise, the thawing process is quite abrupt and lethal damage to the tissues occurs under such situations. Under natural conditions, the plants which have experienced only intercellular freezing, the slower thawing of frost hit plants and plant parts save them from fatal damage. Early morning spray of water onto the surface of frosted plants helps in slower thawing and reduces the level of frost damage to a greater extent in many plant species .

2. Foggers

As described above, the overhead sprinkler system has certain limitations like a large amount of water is needed, a large fraction of applied water does not reach many twigs of the trees but, excessively wets the orchard soil which leads to clogging of soil micro-pores and soil compaction. Such soils need reworking for bringing these back to normal porosity. Further, under many situations, it happens that a significant part of the fruiting wood gets reduced if breakage of limbs occurs due to excessive icing. Foggers are better alternative under these situtations.

Under natural conditions, the foggy nights are usually frost free due to increased air humidity and reduced escape of radiations from an orchard system. Artificial fogging is therefore used for attaining protection against frost, especially hoar frost. The artificially generated fog creates an insulating layer of water vapours which prevents heat loss from the orchard system. Under severe frost conditions, the water vapours from the saturated air release a considerable amount of energy to the orchard system which helps in the prevention of frost damage to the plantations.

The fogging system requires specialized equipment for the generation of fog. High pressure withstanding lines and nozzles are used for the creation of artificial fog. The thickness of the foggy layer is regulated by regulating the water droplet size. The aerosol of fog droplets provides excellent control against frost damage. This system uses considerably a lower amount of water and energy than the sprinkler system. Fogging creates a thinner coating of water droplets at the plant surface which prevents excessive icing and breakage of limbs due to heavy ice load. Automation of fogging system can be done with the help of temperature evolution regression models and S-frost protection guide chart as discussed in Chapter-8. The use of forecasting models and guide chart is essentially required for operating the fogging system under chronic frost-prone situations.

3. Irrigation

Under the Chapter-3 while discussing the orchard energy balance, it has been described that dry soils lose and gain energy easily. The air space present in the dry soil prevents heat storage and therefore creates a more conducive situation for the occurrence of frost. Wet soils, on the other hand, are quite preventive in frost damage especially when the low temperatures spells are lesser in number and duration. Wetting of the soil, often makes it darker and increases the daytime absorption of solar radiation. Further, the wet soil stores a larger amount of solar energy as the water present in the soil possesses better specific heat than dry soil. There are reports which state that the soil water content should be maintained near field capacity under frosty conditions for mitigating its adverse effects. But, usually this has been observed to hold good when the frosty conditions persist only for 4 to 5 nights. Under longer durations of freezing stress, if ice nucleation gets initiated in the soil water, the whole content of water present in the soil freezes very fast resulting in drying and heaving of the soils. Such conditions cause severe water deficit and lead to desiccation of the plants. Added to it, the presence of excess water at the surface aggravates evaporation under windy situations and this drops the soil temperature. This type of situation tends to counter balance the benefits of better radiation absorption of the daytime. The best practice for achieving benefit through this method of frost protection is to wet the dry soils well in advance of the frost event so that the Sun can warm the wet soil during sunny days. By the time the heavy frosty conditions prevail, the soil moisture reaches a level well below the excess. Further, it is unnecessary to wet the soil deeply because most of the daily heat transfer and storage occurs in the top 30 cm, therefore, deep irrigation in winters is simply a wastage of water. Adding excess water to the soil also suffocates the plant roots and increases their vulnerability to the rots. It is therefore advisable that moist conditions should be maintained in the soil instead of wet conditions during winters. Maintaining soil moisture at 50 to 75% of the available water is beneficial for mitigating the adverse frost effects. The frequency of irrigation is generally kept very low in comparison to the summers. It is always better to irrigate on a bright sunny morning than irrigating during the evening or on a non-sunny day. Winter irrigation should be given consideration only when rainfall is lesser than normal. The left-over moisture in the soil from last season's irrigation also reduces the requirement of winter irrigation.

4. Drip Irrigation

Trickle or drip irrigation is a low volume irrigation system that is helpful for frost protection under low to medium intensity frosty situations. Under such situations the water applied through other irrigation freezes on the soil surface and releases some energy into the orchard system. But, under windy situations when evaporation occurs, more energy gets consumed and this leads to damage instead of providing protection. As, drip irrigation system helps in maintaining the soil moist and thus quite useful under such situations. Under heavy and prolonged frosty conditions the use of a drip system is usually not recommended because under such situations the low volume of water present in the tubing of the drip system gets frozen leading to the collapse of the entire irrigation system. If the cost-effectiveness is to be compromised, the use of heated water can give better protection under such situations.

5. Heaters

Heaters are in use for more than 2000 years for the prevention of freeze damage in plants during a frost event [180]. Early in history, the heaters were mostly open fires, however, in recent history the metal containers were used for hosting fire and for better retention of heat which they impart by way of radiation and convection heat to the crop. Later, the growers started the use of simple metal containers that burn heavy oils or old rubber tyres along with sawdust or other farm residues. The fire produced through this method generated oily smoke which was believed to block the radiation losses to the upper atmosphere from the orchard system. But, now it is known that little or no protection is attained by adding smoke particles to the air $_{[180,181]}$. A slight movement of air spoils all the energy accumulation beneath the smoke blanket. Also, the pollution caused due to this method became a serious concern. Therefore, the use of orchard heaters declined slowly. Many countries have now banned the use of heaters for frost protection. In the United States, the growers started the use of "return stack" heaters and clean-burning propane-fuel heaters, the use of which is considered legal in many regions and fuel cost is an issue. Due to the association of a number of problems with the use of heaters, their use in frost protection diminished after the mid-1900s.

6. Wind Machines

Under frosty situations, the cool air pool near the ground and the warmer air gets lifted to the higher elevations, a wind machine is used to mix the warmer air from the upper portions of the inversion layer with the colder air near the ground leading to a rise in temperature of the air in the orchard by a few degrees. Also, the churning of air slows down the freezing process and helps in the mitigation of frost damage. For attaining sufficient frost protection with the use of wind machines, the machines must be turned on when the air temperature in the orchard is still above critical temperatures. It is quite possible that the temperature of the buds may be several degrees below the air temperature due to radiative cooling. Under the situations of a larger difference between air temperature and plant surface temperature, buds may suffer damage even if the air temperature is above freezing temperature. It is better to take the help of 'S- Frost Protection Guide Chart' for deciding the On-Off timings of the wind machine. The use of wind machines for frost protection could not get much popular because most of the wind machines are motor driven and work on fuel. Though, the consumption of fuel is lesser than the heater system but, the high initial cost and tedious maintenance of wind machines are among the main factors due to which the growers shun the use of wind machines for frost protection in their orchards.

7. Orchard Foams

During the mid-nineties, another alternative came into use for frost protection of fruit plantations; in the United States growers started the use of aqueous foam over crops for both freeze and frost protection. Aqueous foam is a good insulator with relatively low thermal conductivity. Gelatin-based foams are commonly used for frost protection as they are more stable, adhesive, biodegradable, long-lasting and serve as a barrier to conductive, convective, and radiative heat transfer. These foams are generally applied in various configurations, directly over the soil surface and spread over a mesh that covers the plants. But, due to their bulkiness, some phototoxic effects, possible environmental hazards and difficulty in removing the foam; this method of frost protection couldn't gain much attention from the growers.

13

Augmentation of Frost Protection Methods

Augmentation strategies are generally used as a way to improve the effectiveness of passive or active frost protection methods. As the intensity of the frost is quite variable and sporadic in nature, simple application of the protection methods is not always helpful in giving adequate protection against freezing stress. There are certain techniques that can be adopted for having enhanced effectiveness of frost protection methods. These techniques are useful especially under situations where localized effects of ecological components are pronounced or where the cost-effectiveness of the protection methods is poor. There are several ways by which we can add to the efficacy of the protection methods by imparting marginal efforts in the main protection method. Some of the augmentation strategies or techniques which can enhance the efficiency of frost protection methods are discussed hereunder:

1. Use of thermal hysteresis for freeze restriction: Freezing injuries in plants may be avoided or reduced by depressing the freezing point of the water present either in the plant system or at the surface or the stored water which is to be used for sprinkling or irrigation. In thermodynamics, thermal hysteresis is a process of change in freezing behaviour of the fluids. De Vries [9] was the first who demonstrated thermal hysteresis. He described it as a condition where the thermal history of a fluid or system defines its behaviour and properties, as it happens between freezing temperature and melting temperature of the water when these temperatures are different. The pure water melts at 0°C, but will not freeze and super-cools to about -40°C in bulk (and even further on surfaces), when no ice nucleators are present [182]. Under the presence of ice nucleators, the melting and freezing point of water is almost the same *i.e.* 0°C; means no thermal hysteresis takes place. When ice nucleator is present together with antifreeze proteins or thermal hysteresis inducing compound, the freezing point of water gets depressed below the melting point (0°C) and thermal hysteresis gets restored. Biological antifreezes having unstable equilibrium show this feature as a primary manifestation of antifreeze activity. The antifreeze proteins or other thermohysteric compounds bind to a reactive plane of ice crystal depending upon their nature and result in inhibition of the growth of ice crystals within a certain temperature range. Groenzin and Schultz [183] illustrated through sum-frequency generation spectroscopy that the surface of ice crystal is highly reactive and the reactivity is influenced greatly by the type of particular face (basal or prism) exposed. This reactivity can be depressed significantly by a **thermohysteric compound** that binds to the reactive face of the ice crystal reducing the number of reactive points available for the binding of water molecules to the ice crystal. This supports the earlier views of Wilson [184], Jorov *et al.* [185] and Madura [186] who opined that the thermohysteric compounds decrease the non-equilibrium freezing of water by directly binding to the surface of an ice crystal (non-basal face), thereby disrupting the normal structure and growth pattern of the ice crystal. It not only inhibits further ice growth but also prevents its re-crystallization and thus imparts thermal hysteresis thereby changing the freezing point of cellular fluids. Certain compounds like alcohols, sugar-alcohols, glycols and antifreeze proteins etc. when added to a cooling fluid, lead to depression in the freezing point due to increased enthalpy of the solution. This in turn reduces the tendency of the system to freeze, and a lower temperature is reached before equilibrium phase transition occurs [187]. The thermal hysteresis inducing compounds thus allows the system to super-cool without ice nucleus formation and preventing the growth or re-crystallization of water molecules.

The thermohysteric property of certain compounds was tried by Sharma [216] for freeze restriction and frost protection in subtropical fruit plant species. Chemical compounds like ethylene glycol, 1-2 propanediol, sodium starch glycolate, ethanol and methanol were applied as a foliar spray for studying their effect on extrinsic and intrinsic freeze initiation and spread in the subtropical fruit plants. The foliar mist was applied just prior to the initiation of the freeze event which was monitored under infrared video thermography during a simulated radiation frost event. The findings of the experiment revealed that all of the applied chemicals influenced the freezing behaviour of all the plant species studied. The effect of these applications on freeze restriction or frost damage prevention was assessed on the basis of electrolyte leakage observed in the freezing exposed leaves. These observations were verified through visualization of the freezing event through infrared video thermography. The highest reduction in relative electrolyte leakage of different species was observed with the mist application of 1-2 propanediol at a minimum concentration of 5%. Application of other chemical compounds like ethylene glycol, ethanol and methanol also produced a significant reduction in electrolyte leakage in the freeze affected leaf tissues. The infusion of these chemicals into sprinkler water considerably reduced the requirement of re-wetting the leaf surface and operational duration of sprinkler or fogging equipment when used for freezing restriction during a frost event. Thus, it was speculated that in the case of natural frosty conditions, only one or two mists of water containing these compounds at a concentration of 5% are sufficient for attaining protection against frost-induced freezing in plant species that experience extrinsic ice nucleation. For enhanced efficacy of the application, the automation of the spray or fogging equipment can be done as per 'S- frost protection chart', the use of which has been detailed in Chapter-8. The excessive use of water in foliar sprays and water stagnation of the orchard soil gets considerably reduced by agumatation of the sprinkler or fogging system of active frost protection with thermohysteric compounds.

Like the other methods of frost protection, the use of thermohysteric compounds was also found to have certain limitations. The foliar application of thermal hysteresis inducing compounds was not observed effective in the case of plant species that are highly frost-sensitive or experience intrinsic ice nucleation first to the extrinsic nucleation. Also, prolonged benefits of their application were not observed. Spray application of these chemicals through sprinkler or foggers methods was found effective only against the ongoing frost event. Significant residual benefits of their application were not observed for the next frosty night. Afresh foliar application of these chemicals is required for achieving a significant reduction in frost damage. Further, for effective restriction of freezing damage, the application of the thermal hysteresis inducing compounds is required just prior to the ice nucleation event. Under natural conditions we can't guess the exact time of ice nucleation or freeze initiation, hence the spray or mist applications are required either to be continued throughout the night or to be applied intermittently with the aid of automated spray equipment which could deliver the formulation at specified intervals of time. Added to these, the size of the trees is also a big problem. Most of the commercial subtropical fruit orchards happen to be more than 5 meters in height. For having an effective application of the chemical formulations, it requires the machinery of heavy capacity for making proper overhead spray.

Horticulturally, mist or foliar application is the most convenient way of applying the desired chemical formulation to a target plant. But, as discussed above there are certain limitations in the use of thermohysteric compounds through spray application, alternative ways were explored for having prolonged benefits of thermohysteric compound application. Experiments were conducted under the controlled growing conditions to ingress these compounds into the plant system either through drenching the soil or by the placement of saturated hydrogel beads in the effective root zone or by band application by ringing the stem or shoot bark. The freezing stress responses of the plants were studied in relation to these different ingression techniques. The experiments concluded that placement of 1, 2 propanediol (5% solution) saturated hydrogel beads in the root zone of the plants restricted freezing damage in most of the subtropical fruit species to a longer duration than other chemicals under study. The effect of other chemicals and methods was also significant. These findings are yet to be confirmed under open environment orchard conditions where the size of the plant rhizosphere and canopy are larger and non-consistent frost intensities prevail.

2. Reducing Surface wetness: Though frost is a phenomenon of natural occurrence, the extent of damage to a particular crop species depends entirely on the level of freezing experienced by that species. Plant parts usually freeze when they cannot avoid nucleation and growth of ice crystal within their system during a frost-induced freezing. The speed of ice crystal growth and its subsequent propagation into the cell interspaces is believed to have been facilitated by the presence of free water and depends directly on its amount present at the surface. This presence of free water at the leaf surface gets created during a natural frost freeze event as a result of condensation just before the frost point is achieved. More condensation of water vapours onto the leaf surface occurs under the saturated atmospheric conditions before the natural freezing event[14].

To develop a preventive mechanism for the restriction of freezing under a frost event, a study was conducted by Sharma [126] to reduce the amount of free water at the phylloplane with the use of hydrophobic compounds. During a freezing event, free water at the leaf surface was reduced by foliar application of hydrophobic chemicals like Aluminum tris- (o-ethyl phosphonate), Kaolin, C3- 4-hydroxybutanoic Acid Lactone, tri-ethylsilane, dimethylpolysiloxane and vegetable oil emulsion. The effect of these chemicals was studied on the freezing behaviour of Galgal, Guava, Jackfruit, Jamun, Karonda, Lime, Litchi, Loquat, Mango and Papaya under simulated frost conditions. The freezing damage to these plant species was assessed through relative electrolyte leakage studies. The experimental findings concluded that the highest reduction in relative electrolyte leakage with the application of hydrophobic was recorded in the case of mango, guava, and litchi. These fruit plant species otherwise suffered heavy freeze damage when exposed to freezing stress with wet leaf surface without any application of hydrophobic compound even under natural frost conditions [126]. These species experience extrinsic ice nucleation which is the primary cause of freeze propagation across the plant organs. The ice nucleus formed at the leaf surface grows fast under the presence of free water at the leaf surface and propagates into the plant system through the stomatal openings, hydathodes and wounds. When hydrophobic compounds were applied, surface water was reduced; the extrinsic ice nucleation was restricted. In the case of most of the frost-sensitive fruit crop species like mango, guava and litchi the effect of foliar application of all hydrophobic compounds was significant and the lowest damage was recorded with the application of 5% dimethylpolysiloxane. In the case of plant species which are highly frost sensitive and experience faster internal freezing, the use of these compounds was found ineffective. In species like papaya and jackfruit, the intrinsic ice nucleation occurred much faster than extrinsic ice nucleation, so by the time the hydrophobic compounds could show their effect, internal inter or intra cellular freezing of water occured in the leaves. The effect of hydrophobic compounds also could not be differentiated in the case of high frost-tolerant species loquat and citrus.

Overall, we can say that the foliar spray of the hydrophobic compound has been found effective in mitigation of the adverse frost effects only in the case of plant species that experience extrinsic ice nucleation as a prominent cause of freeze initiation and spread. In the case of highly frost-sensitive plant species or the plant species in which intrinsic ice nucleation occurs first, least benefits due of hydrophobic compound spray application was observed. Effective restricting in ice nucleation and growth with the application of acrylic polymers and hydrophobic compounds has also been reported by Glenn *et al.* [188], Sapienza [189], Fuller *et al.* [190]. Wiesniwski *et al.* [191] mentioned that Citrus species can be super-cooled up to –6°C, despite the presence of a strong nucleation factor with the help of hydrophobic barriers.

3. Energy Conservation within the Orchard system

Solar energy is the primary source of heat that imparts the highest energy to an orchard system. As discussed earlier in Chapter-3, frost is an energy related phenomenon and for its occurrence, the net energy balance of the orchard is required to go negative. On a cool, clear and calm night; there occurs loss of energy in the form of infrared long wave radiations from the earth surface and other components of the ecosystem. For frost to occur, the loss of radiation energy during the night must exceed the gain acquired during the daytime. Thus, the level of net radiations of the orchard ecosystem is an important factor that determines the prevalence of frosty conditions. The efficacy of protection methods gets enhanced several times with a slight positivity of the orchard energy level. The longer the positivity of the net radiation is maintained during a frost event, the lesser is the need for employing the protection method and the lesser will be the damage caused due to frost-induced freezing. Thus, any cultural practice or other orchard management measures which can reduce the radiation losses during the night or can enhance the solar light harvesting into the orchard ecosystem during the day, can be of great help for reducing the chances of frost damage.

There are many orchard management practices that can help the growers to maintain the positive energy balance of their orchard. Described below are some of the practices which can be adopted for energy conservation at the orchard level:

a. **Removal of Weeds and Cover Crops**: Weeds and cover crops are the biggest barriers between earth surface and solar energy; keeping the orchard clean helps the soil to receive more energy during the daytime. Therefore, weeds and cover crops if present, should be removed prior to winters for enhancing the solar radiation absorption by the orchard soil. If the orchard floor is managed under permanent sod, the grasses/sod present shouldn't be taller than 5cm during winters. The taller orchard grasses may cause a difference of as high as 2^0C compared to no-grassed orchard floors during a radiation frost event. Clean inter-row spaces give better energy balance for the orchard than the permanent sod.

b. **Avoiding Tillage**: Loose soils conserve less solar energy in comparison to the non-tilled ones; therefore, tillage should be avoided during winters. For better orchard energy balance, the work of tree basin preparation should be completed well before the onset of winter. Cultivation of soil creates larger air space; as the air has poor specific heat, it tends to transfer and store lesser heat than the compact soil. If the soil is already cultivated, its heat transfer and storage capacity can be considerably increased by levelling it.

c. **Removing FYM from orchard surface**: Farm Yard manure (FYM) possesses very low specific heat therefore, should not be left lying on the orchard floor. It should be mixed well with the soil just prior to winters. Early winter mixing of manure is beneficial from the nutritional point of view, also.

d. **Keeping the soil moist**: Under non-windy conditions, the addition of water to the orchard eco-system maintains the positivity of net radiations by increasing the latent heat level within the orchard. Water stores a larger amount of solar energy in comparison to dry soil. Thus, keeping the soil moist can help in a great way for harnessing daytime solar energy and mitigating the devastating effects of frost.

e. **Harnessing and storage solar energy**: If there is a larger water storage structure in the orchard it should not be covered during the winters. Solar energy should be allowed to fall directly on the water so that more and more energy could be stored in it. Larger stones or buildings etc. also store a good amount of solar energy and help in maintaining the positive energy balance of the orchard during the night through irradiance.

f. **Thatch Cover:** Thatching is an age-old practice to protect plants from frost damage. Conical thatch cover over the small plants gives excellent protection than the flat thatch covers. Conical thatch covers encase the plants better than flat roof thatch covers and prevent heat loss in a better way. A slight opening on the Southern side of the conical thatches should be kept for the light and air entrance to the plants. Over the nurseries only flat thatch covers are possible; under such situations, the height of flat thatch should be not more than 30cm above the head of plants otherwise, free passage for the movement of cold air get created above the nursery plants and huge damage sometimes occurs.

g. **Overhead Plastic Net Cover:** Foliage of the sensitive plants can also be protected to some extent by covering with plastic nets of fine meshing. Two to three layers of plastic net are spread over the canopy of the plants or nurseries; it gives good protection under low to medium frosty conditions. Normally, shade net of 60% shading or Woven and spun-bonded polypropylene plastics can be used in single or two layers for this purpose. These covers reduce the loss of radiation heat from the soil & plant surfaces and maintain the warmer conditions beneath for a longer time. Depending upon the material used and the meshing per cent of these covers, the net may help in keeping the temperature 1 to 2^0C above the surroundings. Clear plastic sheets also give good results but care is required to be taken for the aeration of the plants. Excessive condensation underneath these covers usually invites many pathological problems. Also, the distance between the plastic film cover and the plant head matters for the efficacy of this protection method.

h. **Soil cover:** In order to conserve daytime energy in the soil, it is recommended to cover the soil surface with transparent plastic film. Wetting of soil prior to applying soil cover sheet provides much better results. Covering the soil surface with vegetative refuse should be avoided in winters as it hinders the solar energy harnessing into the soil surface. Absorbance of high amount of solar energy by bare soil is otherwise very helpful for maintaining the positive energy balance of the orchard.

4. Proper training & pruning

It is a general principle that the surfaces which are flat/facing the sky irradiates larger energy at a much faster rate during a clear calm night than an inclined surface. The same is true for the branches and leaves of the plants. Under frost-prone situations, to minimize the radiation losses and for maintaining a positive energy balance of the orchard for a longer duration, the tree branches should be trained at about 60°-75° with the central axis of the tree so that the radiation losses and frost damage could be minimized.

The upper canopy of the evergreen plant species acts as an umbrella for the protection of the lower canopy against frost and low-temperature injury. Removal or thinning of this canopy portion prior to winters is thus not advisable. The pruning operation in evergreen tree species should either be performed after harvest or during late winter. This also enables the growers to retain resource wood which will remain available for growth and productivity in case of hazardous frost event. Whenever pruning operation is being done, it should be kept in mind that the lower canopy of the tree should be so pruned that there remain no branches touching the ground. The canopy skirt should be kept around one meter above the ground in order to facilitate the penetration of solar radiations below the canopy up to the earth surface. It helps in warming the area underneath the canopy during the daytime. In the case of frost damage to the upper tree canopy, pruning of dead wood should be done only during late winter or spring season because the dead foliage also provide frost protection to the lower canopy. If during pruning larger limbs are required to be removed, the cut ends of these limbs should be properly covered or waxed.

5. Waxing or Covering of Wounds

As the evergreen subtropical species remain physiologically active during winters, their pruning of just prior to winters should be avoided. If pruning is done during winter or early winter, the wounds at the cut ends of the branches remain live and oozy. The exudation of sap from these wounds serves as a source of free water at the plant surface and facilitates the ice nucleation and spread during a frost event. An unhealed pruned wound may serve as the entry points for the surface ice crystals to grow into the plant's vascular system leading to lethal damage to the plants under severe frosty conditions. The whole plant from top to roots may die under such circumstances. Even if, the plants are pruned quite early during autumn then also, the cut ends of the branches should be waxed or covered with poly-sheet before winter so as to prevent the entry of the extrinsically nucleated ice into the plant system and preventing lethal damage. The proper procedure of wound waxing and covering is described in the next chapter.

6. Use of Trunk wraps

During winters, after sunset the temperature of the tree trunk surface drops very fast; this usually kills the stem tissues and leaves the bark cracked and dry. In many fruit species like *Citrus*, aonla, jamun, harar etc., there occurs heavy damage to the tree trunk in the form of tree bark splitting under prolonged frosty situations. Such plants can be protected from frost damage to a greater extent by providing a trunk wrap with non-hygroscopic, thermo-insulating material. The trunk wraps prevent heat loss from plant's trunk surface besides harnessing better sun energy during daytime, if dark coloured warps are used. Such wraps also protect the plants from wild animals, which otherwise eat the stem bark during winters due to food scarcity. Trunk wraps also protect plants from sunscald which usually damage the bare trunk of high canopy trees. Sunscald occurs mainly on the southern or South-West facing part of the tree trunk. For the protection of the plants from frost, the trunk wraps should be applied during autumn or before the onset of the frosty winters. These wraps should be removed as early as the threat of frost is gone; otherwise, they act as hiding space for the pests.

7. Trunk Whitewash or Paints

The whitewashing of the trunk is a very important pre-winter operation for the fruit orchards. This practice helps the plants to maintain uniform stem temperature, prevent sunburn, eliminate the hiding or hibernating pockets of the insects under the bark and add to the ambience of the orchard. White colour being a poor absorber and radiator of radiations reduces the chances of excessive radiation absorption during the day and excessive radiation loss during the night. Whitewashing therefore, reduces the cracking of bark by buffering the wider temperature fluctuations between days and nights of the winter. As, the frost damage reduces the volume of the upper canopy and the main stem and branches get exposed to the direct sun in the summers. If not whitewashed properly, these branches and stem suffer heavy damage due to sunburn. Whitewash is also helpful in the protection of plants from pathogens and diseases because it fills the crevices of the bark and eliminates the hibernating or hiding sites of pests and pathogens. Therefore, good orchardists always whitewash the trunk and main branches of the orchard trees before the onset of winter.

Whitewash solution can be prepared from white latex paint by mixing it with water in equal quantity. If the whole crown of the foliage is lost, spray application of this emulsion may be done by further diluting it to spray-able concentration. Alternatively, a whitewash solution may also be prepared from slaked lime (Calcium hydroxide) by mixing it with water and other components in the following proportion [192].

Whitewash solution = Base + Water + Fungicide + Anti animal component + Fixing component

Whitewash solution=Lime-2kg + water-10lt+ Copper Sulphate-200g + Carbolic Acid-15g (1table spoon)+Gum/Fevicol DDL-40g

The ingredients are required to be mixed thoroughly before applying to the plant trunk surfaces. Before whitewashing the trunk of the trees, the dead bark should be removed by rubbing the stem with a jute mat.

8. Pesticide or Herbicide sprays

Many of the pesticide or herbicide formulations contain solvents or adjuvants like alcohol, glycols, benzene and acetones etc. which possess thermohysteric properties. Spray application of such formulations sometimes help in mitigating the adverse effects of frost especially those which occurs to blossom during spring frost. The thermohysteric additive constituents of the pesticide formulations when sprayed for control of insect-pests get mixed with the dew/ water at the blossom or plant surface and restrict its freezing process, save floral parts from freezing under medium to mild freezing situations of the late-blooming crops during spring frost. The herbicide sprays, on the other hand, make soil bare, capable of absorbing high solar energy during daytime and thus, improve orchard energy balance. Besides, the thermohysteric constituents of the herbicidal formations restrict freezing of water at soil surface.

14

Growth Restoration in Frost Affected Orchards

If frost has hit an orchard during winters, it does not always imply that permanent loss has occurred. Plants do possess natural protective mechanisms due to which they survive the stress to a variable extent depending upon their nature and organ being exposed to the stress. If the plant's vascular system is not affected, there remains certain tissues and organs which successfully thaw back to normalcy and regenerate partially or fully with repetitive efforts of growth restoration. The growth restoration is a tedious work in severely affected orchards. When orchards are damaged to a greater extent there come a challenge of removing the damaged wood and reinvigorating the remaining portion of the plant with the efforts for saving the fruiting wood to the extent maximum possible so as to reduce the economic losses. Following are some of the important practices which can be followed for faster recovery of frost hit orchards:

Regenerative Pruning

As and when the frost stress is over, the growers immediately try to remove the damaged portion of the plant. It is generally a wrong practice; the dead portion of the plant should be retained till the complete picture of frost damage gets apparent in the spring. There are many benefits which are linked to this decision. It allows sufficient time for the new growth to appear and the extent of damage to be clearly visible. It reduces the chances of unnecessary removal of healthy wood. Apart from this, the dead material serves as a protective cover for the living portion of the plant against the upcoming late winter or early spring frost events. If we prune out the dead material before another frost event, there are chances that more portion of the plant gets exposed and killed due to frost. Further, if late frost will occur, the cut ends of the branches will serve as the entry points for the progression of the freeze through the vascular system up to root and shoot apices and may lead to the complete death of the plant. Most of the subtropical fruit plants survive frost stress if a freeze does not get entry into the plant system and the vascular system is intact and alive. Without entry of extrinsic ice crystals into the vascular system it remains vibrant and live owing to its solute concentration and fluidic movement and due to this reason, the woody portion or hardier growth of the plants do not suffer serious damage until unless the frost event is of very high intensity. The damage which occurs to foliage usually sheds out and new growth emerges in the spring in many of the subtropical woody perennials. But, if leaves remain there at twigs, it is an indication that the plant has suffered seriously due to rapid freezing. The branches or main stem which has suffered such type of damage can be identified from the cracks or splits in the bark. All these symptoms of damage become visible only after the resumption of the growth in the spring. Therefore, it is very necessary to wait before going for the removal of frost damaged portion of the trees.

With the onset of the spring season, the young and undamaged buds break early however, the buds lying in the older parts of the plant take a longer time to emerge as these buds need more warmth and stimuli for emergence. The extent of damage, under such situations may be accessed by making a cut 1 to 2cm above the expected live bud, if the vascular region is alive and greenish-white in colour the bud will emerge but if it is brown or dark brown we should go down further to the lower bud to make the next cut. This bud inspection mechanism should also be followed before going for the first pruning of the damaged branches of the plant and it should be repeated till we get live bud at a position on the branch from where we can have a new branch in the right direction as per the requirement of the tree canopy. Such type of careful pruning of branches helps in directing the new growth in the proper direction right from begining and eliminate the need for additional pruning at the later stages.

Frost sometimes kills some of the plants up to the ground level; if it is the case, we have to wait and watch if the plant is growing back or not. To confirm this, it is needed to make a clean sharp cut about 15cm above the ground level for inducing the growth of latent epicormic buds. Some of the plants come up with in 2 to 3 weeks but some seriously damaged ones may take 3 to 4 months for regaining growth. Irrigation and exposing the soil around the trunk to Sun may help in a faster resumption of growth in such cases. The herbaceous and succulent plants like papaya suffer complete loss of foliage and the stem gives a healthy appearance even after suffering heavy internal damage. Such papaya plants mostly bear dry leaves on the leaf stalk and some young alive leaves at the stem apex. As the spring comes and the temperature starts rising such plants get invaded by rotting microorganisms (bacterial and/or fungal) and collapse soon. These plants rarely survive for two or three months. Early diagnosis of such plants can be done in the spring by making a small incision just above the collar region of the plants and extracting some internal tissue with some needle or other instrument. If the internal stem tissue is damaged it will give a foul smell, if it is not damaged seriously it will smell normal.

Further, pruning of the frost damaged plants is essential also for their prevention from invading pathogens and wood decaying organisms which otherwise slow down the recovery. Therefore, as described above, it is not wise to rush to prune. Delay pruning until danger of the frost has passed. It should be remembered that pruning is tough on plants; the more we prune, the more is the energy is lost to the wound healing process of the plant. It costs the plant new growth and flowering. While doing pruning, all cuts should be made into living wood at crotches with leaving no branch stubs. All cuts should be smoothly finished. The pruning tools should be sterilized before and after pruning a tree. For disinfecting the tools isopropyl alcohol (70%) or sodium hypochlorite solution (10% bleach solution) or home scale '*Lyzol*' solution can be used [193].

The pruning operation should be restricted only up to small branches of diameter not more than 3cm; cutting of bigger branches should be avoided during late winter or spring season, these remain oozy. All the cut ends should be treated with Bordeaux paste immediately after pruning. Every care therefore, should be taken for healing of the wounds; otherwise when the plant resumes active growth, oozing of the sap from the cut ends starts and it serve as the substrate to the pathogens. It is advised to apply clay-cotton mix at the larger cut ends. Clay-cotton mix can be prepared by soaking clay in water then kneading the cotton in it. Apply the kneaded clay-cotton mix onto the cut ends, it will absorb the excess sap and will prevent the decay of the wood by various micro-organisms. Bigger wounds should be waxed or covered with a polythene sheet. It has been seen that the plants which were pruned heavily or having bigger broken branches get affected badly by the frost in the ensuing winter season. The open exposed cut ends of the branches serve as the entry points for the ice crystal to grow into the plant system. This leads to freezing of the vascular sap and complete death of the plants. Therefore, pruning prudently is very essential in the case of evergreen plants under frost-prone situations.

Pic. 9: Covering of branch cut-end-wounds with clay-cotton mix and wrapping these with polythene sheet to avoid damage to plant vascualr system

Foliar Application of Growth Hormones

Plant hormones play important functions in the control of the growth and development of all plant species. Abscisic acid (ABA) has its specific role in plant responses to stress. Upon perceiving the signals of low-temperature stress, the plants tend to accumulate ABA so as to coupe up with stress. In deciduous woody species, the levels of ABA tends to increase with the onset of winters and declines progressively after the fulfilment of the chilling requirement of the plants, such reports on ABA transformations are scanty in the case of evergreen subtropical woody species. Though enough evidences are not available to decipher the role of ABA acclimation of evergreen subtropical species, still there are evidences which indicate a relatively higher level of ABA than other hormones, during exposure of plants to cold stress. It has been speculated that resumption of growth at a normal pace after intense winters enhances the synthesis of growth promoting hormones which happen essentially in synchronization with the cessation of winters and for the negation of the inhibitory effects of ABA.

The plants, which suffer heavily due to winter frost usually show delayed response of growth resumption due to weak action of the growth promoting hormones. In such frost hit plants, the bud-takes may take few days to months for resumption of normal vegetative and reproductive growth due to reduced functional leaf proportion. The accumulation of growth promoting substances and negation of growth inhibiting factors take a long time in these plants. The exogenous application of growth promoting hormones therefore results into good growth restoration depending upon the remains of live-wood and foliage over these plants. As cytokinins have a high antagonistic effect on ABA and a synergistic effect on shoot proliferation & development; it was tried for growth restoration of frost hit mango orchards by Sharma [194]. The foliar application of benzyladenine (BA) at a rate of 20ppm just at the onset of spring, followed by 1% urea spray after fifteen days produced better and fast shoot regeneration and floral bud opening in the frost hit mango trees. Better fruit set and yield were also obtained in these trees than the untreated ones.

Manure & Fertilizer Application

There are general recommendations for pre or early winter application of manures, phosphatic and potassic fertilizers in subtropical fruit orchards; additional application of these manures and fertilizers is not therefore required in the spring. However, the scheduled nitrogen application for the spring should be applied in the frost hit orchard as early as the new flush starts emerging. Its application should be done very carefully because the reduced foliage of the frost hit orchards can use only a limited amount of nitrogen at once. Therefore, the excess availability of nitrogen may enhance the emergence of water shoots from the lower juvenile portion of the trees, which is not desired. The emergence of water shoots delays the sprouting of desired vegetative and reproductive growth on a frost affected plant. Despite precautions, if a water shoot emerges, it should be removed at the earliest to avoid biomass loss through its growth. Once the plants regain their stature and growth, they need nitrogen more than normal for recouping their lost organs and tissues. Therefore, split doses of nitrogen application give better restoration of the frost affected orchards.

The orchards with poor nutrition level suffer heavy frost damage therefore, regular leaf sampling of fruit orchards is essential especially in the frost-prone agro-ecological settings. The deficiencies of micro-nutrient such as zinc, manganese, copper, iron etc. are more probable under such situations, Therefore, close monitoring of nutrient deficiencies should be carried out for mitigating the adverse effects of frost.

Irrigation

Irrigation of the frost affected orchards early in spring usually leads to root suffocation and proliferation of root diseases because, after the damage to the foliage due to frost, the balance between the roots and foliage get disturbed. The presence of a reduced number of transpiring leaves greatly reduces the need for water. The water requirement of such plants becomes as low as that of a young and healthy plant. Full irrigation of the plants, as these were receiving before damage, may lead to slower recovery due to root suffocation. Prolonged persistence of excess moisture in the soil leads to the rotting of roots. It is therefore, advised to give only light and frequent irrigations to the frost hit orchards until the plant regains their foliage and vigour.

Whitewashing the tree trunk: In heavily frost damaged orchards the foliage becomes sparse. The whitewashing of the trunk and main branches is therefore, essentially required in such orchards. It is a standard orchard management practices to whitewash the trunk of the orchard trees just prior to winters, if not done in the winters, it can be done during the early spring in the frost hit orchards. This simple practice not only helps the trees to maintain uniform stem temperature but also eliminate the hiding and hibernating pockets of the insects under the bark. It protects the frost affected, sparse foliaged trees from sunburn during the summer and also adds to the ambience of the orchard. Whitewash reduces the chances of bark cracking in partially damaged stems and branches under the situations of wider differences in the day and night temperatures.

Good orchardists therefore, always whitewash the trunk and main branches of the trees before the onset of winter. The formulation and preparation method of whitewash solutions/paints is already discussed in Chapter-12.

15

Frost Impact Mitigation The Way Forward

Global climate change has increased the weather variability across all geographical settings. The transitional regions between the tropics and temperate region are expected to buffer the changes comparatively for a extended duration of time. But, inconsistency of the winter weather patterns is predicted to increase and therefore the vulnerability of the crops to frost also expected to increase due to the oscillation of the plants between acclimation and de-acclimation processes. Thus, crop losses are to be more pronounced within the subtropical regions because of the repeated exposure of plants to frost induced freezing stress. This risk will become more substantial under situations where exposure to freezing temperature will combine with the inherent susceptibility of the plants to low-temperature stress. Concerted efforts for developing effective frost impact mitigation strategies are therefore imperative.

Furtherance of elaborative research on the interactive role of assorted eco-physiological factors related to frost occurrence and endurance under specific geographic settings is needed for the economic survival of the subtropical orchards. Under the lower Himalayan subtropical conditions, beside other factors, the topographical and altitudinal factors have considerable touching on the occurrence and endurance of frosty situations. Therefore, the micro-level analysis and mapping of frost patterns, intensity and duration is required for effective planning of horticultural development in this region. Quantification of the contribution by soil, water and vegetation factors together with microclimate and spatial structure interrelations for different fruit crops will prove instrumental in relieving the fruit plantation from frost induced embolism.

Apart from direct freezing damage, altered metabolism, growth and development are the typical responses of plants that require to be precisely understood especially just in the case of woody tree species of subtropics. The regulatory factors like stress sensors, signaling pathways comprising protein-protein interactions, transcription factors, promoters and ultimate output of defensive proteins and metabolites need proper elucidation for developing frost endurance mechanisms. Role of antioxidative stress related defence genes capable of inducing changes in plant biochemical mechanisms for frost induced oxidative stress need detailed studies in this case. Abscissic acid is the prime hormone related to any type of stress in plants, its role in transcriptional and post-transcriptional regulation, adaptation of cellular functions to the low-temperature environment is yet to be elaborated for many of the commercially important fruit tree species.

In a quest for frost tolerant germplasm of important horticultural crops, the lower Himalayan zone or other cooler regions of the substropics could be explored for screening of untamed native species/strains/varieties having a good degree of tolerance against this natural calamity. Elite selections should be conserved for inclusion in future crop breeding programmes for developing frost tolerant germplasm of commercial importance. There is a considerable advancement within the biotechnological approaches as an alternate to time and money, the major limiting factors of traditional crop breeding programmes associated with abiotic stress resilience development. The discoveries regarding the involvement of single gene (one gene association as an osmoprotectant, detoxifying enzyme, transporter protein genes etc.) or multiple genes (biomolecule synthesis, regulatory transcription factor and signal transduction genes etc.) associated traits are now opening new possibilities for devising solutions, at the cellular and sub-cellular level, against the low-temperature stress. Genetic engineering and genome editing tools are new hopes for the transfer of genes among different species and modify the target genome in a very site and crop specific manner for cold stress-responsive mechanisms.

The growers today don't feel much enthusiastic about adopting currently available frost protection methods, perhaps they have doubts regarding the effectiveness of these methods or the capital cost associated with the equipments and consumables doesn't fit into the orchard economics. Therefore, there is urgent need for developing high precision machines having low capital and operational costs. Equipments with improved configuration and automation capabilities are needed to be developed for zones of variable frost intensity and duration. Fruit orchards usually comprise trees having larger canopies with the variable orientation of branches. The machines are therefore required to be developed with multiple programming modules. This can enable fine-tuned automation of the frost equipments for the economical viability of the orchards.

One amongst the most effective solutions for the endurance of the frost stress is through gaining knowledge on its genesis, progression and crop damage mechanism involved. The knowledge on these aspects can enhance the capacity of the growers to modulate the operation of their frost protection method or equipment as per the requirement of the upcoming or ongoing frosty situation in their orchard. To enhance their experiential understanding of these concepts, it is imperative to put intensive efforts into bringing them up-to-date regarding recent advances in frost protection research and the ways to achieve frost endurance for their subtropical fruit orchards. Added to that on-farm demonstrations and trials for collaborative research initiatives are required so that scientists might make recommendations more specific to local weather patterns, topography and the germplasm being cultivated. Self-awareness is also required to be strengthened among the farmers for monitoring daily changes in weather parameters, frost intensity and crop damage levels. Regular record keeping at individual orchard level can help in a great way for improving the efficacy of frost protection equipment and strategic mangement of frosty situtations in the future. Further, the data sharing through new-age technologies like data collection, mobiles phone applications or data pooling at clouds can take the frost related research to new horizon where crop, weather and micro-scale agro-ecological data can be used for formulating artificial inteligence based solutions for the benefit of the frost affected farming community across the globe.

Botanical Names of the Plants

S.No.	Common Name	Scientific name
1.	Aloe	*Aloe vera*
2.	Aonla	*Emblica officinalis*
3.	Apple	*Malus x domestica*
4.	Apricot	*Prunus armeniaca*
5.	Asparagus	*Asparagus officinalis*
6.	Avocado	*Persea americana*
7.	Banana	*Musa paradisiaca*
8.	Beans	*Phaseolus vulgaris*
9.	Beets	*Beta vulgaris*
10.	Ber	*Ziziphus mauritiana*
11.	Berries	*Rubus* spp.
12.	Bil/ Bael	*Aegel marmelos*
13.	Broccoli	*Brassica oleracea* var. italica
14.	Brussels sprouts	*Brassica oleracea* var. gemmifera
15.	Carrots	*Daucus carota*
16.	Cauniflower	*Brassica oleracea* var. botrytis
17.	Celery	*Apium graveolens*
18.	Citrus (mandarin)	*Citrus reticulata*
19.	Cranberries	*Vaccinium* Spp.
20.	Cucumber	*Cucumis sativus*
21.	Custard Apple	*Anona squamosa*
22.	Dates	*Phoenix dactylefera*
23.	Egg plant	*Solanum melongena*
24.	Galgal	*Citrus pseudolimon*
25.	Grape Fruit	*Citrus paradisi*

S.No.	Common Name	Scientific name
26.	Grapes	*Vitis vinifera*
27.	Guava	*Psidium guajava*
28.	Harar	*Terminalia chebula*
29.	Jackfruit	*Artocarpus heterophyllus*
30.	Jamun	*Syzygium cumini*
31.	Kale	*Brassica oleracea var. sabellica*
32.	Kinnow (mandarin)	*Citrus reticulata*
33.	Kohlrabi	*Brassica oleracea* (Gongylodes Group)
34.	Lettuce	*Lactuca sativa*
35.	Lime	*Citrus aurantifola*
36.	Litchi	*Litchi chinensis*
37.	Longan	*Dimocarpus longan*
38.	Loquat	*Eriobotrya japonica*
39.	Mango	*Mangifera indica*
40.	Okra	*Abelmoschus esculentus*
41.	Onion	*Allium cepa*
42.	Oranges	*Citrus sinensis*
43.	Papaya	*Carica papaya*
44.	Parsley	*Petroselinum crispum*
45.	Parsnip	*Pastinaca sativa*
46.	Passion Fruit	*Passiflora edulis*
47.	Peach	*Prunus persica*
48.	Pear	*Pyrus communis*
49.	Peas	*Pisum sativum*
50.	Persimon	*Diospyros kaki*
51.	Phalsa	*Grewia asiatica*
52.	Plum	*Prunus salicina*
53.	Potato	*Solanum tuberosum*
54.	Radish	*Raphanus sativus*
55.	Raspberry	*Rubus ellipticus*
56.	Rutabaga	*Brassica napobrassica*

S.No.	Common Name	Scientific name
57.	Spinach	*Spinacia oleracea*
58.	Strawberry	*Fragaria x ananasa*
59.	Summer Squash	*Cucurbita spp.*
60.	Sweet Pepper	*Capsicum annum*
61.	Sweet Potato	*Ipomea batatas*
62.	Tomato	*Solanum lycopersicum*
63.	Turnip	*Brassica rapa* subsp. *rapa*
64.	Winter Squash	*Cucurbita* spp.

Colour Plates

Chapter 1: Frost and Subtropical Fruit Growing – An Overview

Pic. 2: Frost Affected Mango Orchard

Chapter 4: Mechanism of Frost Damage

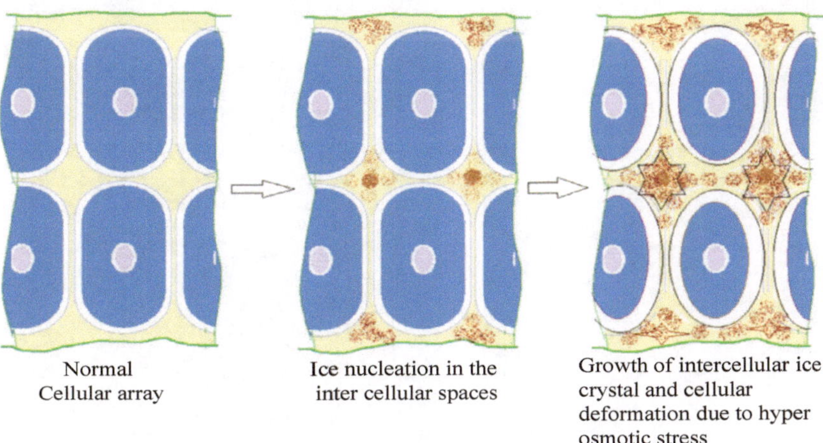

Fig. 3: Pictorial representation of intercellular freezing and hyper-osmotic plasmolysis during a frost induced freezing event

Chapter 7: Frost Damage Symptoms

Pic. 4: Frost damaged Papaya

Pic. 5: Progression of frost induced freeze from bud to vaseular system in mango

Pic. 6: Frost induced stem rot & defoliation of Papaya

Pic. 7: Bark cracking of Aonla due to frost injury

Pic. 8: Frost damaged mango leaf (image developed through drying and rehydration method

Chapter 14: Growth Restoration in Frost Affected Orchards

Pic. 9: Covering of branch cut-end-wounds with clay-cotton mix and wrapping these with polythene sheet to avoid damage to plant vascualr system

References

1. Koppen W. 1940. Köppen climate classification. https://www.britannica.com/science/Koppen-climate-classification
2. Anonymous 2010. http//www.business.gov.in/agriculture/ currentscenario.
3. Rigby JR and Porporato A. 2008. Spring frost risk in a changing climate. *Geophys. Res. Lett.* 351215. DOI: https://doi.org/10.1029/2008GL033955
4. Hogg WH. 1950. Frequency of radiation and wind frosts during spring in Kent. *Meteorological Magazine*, 79: 42–49.
5. Hogg WH. 1971. Spring frosts. *Agriculture*, 78: 28–31.
6. Lawrence, EN. 1952. Frost investigation. *Meteorological Magazine*, 81: 65–74.
7. Cunha JM. 1952. Contribuição para o estudo do problema das geadas em Portugal. [in Portuguese] Relatório final do Curso de Engenheiro Agrónomo. I.S.A., Lisbon.
8. Snyder RL and de Melo-Abreu JP. 2005. Frost protection: fundamentals, practice, and economics (Vol.1). Environment and Natural Resources Series 10. FAO ENRS. http://www.fao.org/3/a-y7223e.pdf
9. De Vries H. 1877. Eine Methode zur Analyse der Turgorkraft. Jahrb. Wiss. Bot. 14, 427–601.
10. Moor H. 1964. Die Gefrier-Fixation lebender Zellen und ihre Anwendung in der Elektronenmikroskopie (Freeze-fixation of living cells and its application in electron microscopy). Z Zellforsch 62:546–580.
11. Levitt J. 1980. Responses of plants to environmental stresses, Vol. 1. New York: Academic Press.
12. Pescod DA. 1965.Method of forming dew on plants under controlled conditions. J. Agril. Engg. Res. 10: 3318-332.
13. Siminovitch D, Singh J and de la Roche IA. 1978. Freezing behaviour of free protoplasts of winter rye. *Cryobiology,* 15: 205-213.
14. Franks F. 1985. Biophysics and biochemistry at low temperatures. Cambridge: Cambridge University Press.
15. Hsieh TH, Lee JT, Yang PT, Chiu LH, Charng YY, Wang YC and Chan MT. 2004. Heterology expression of the Arabidopsis C-repeat/ dehydration response element binding factor 1 gene confers elevated tolerance to chilling and oxidative stresses in transgenic tomato. Plant Physiol.135: 1145–1155.
16. Zhu JH, Dong CH and Zhu JK. 2007. Interplay between cold-responsive gene regulation, metabolism and RNA processing during plant cold acclimation. Curr. Opin. Plant Biol.10: 290–295.
17. Uemura M and Steponkus PL. 1999. Cold acclimation in plants: relationship between the lipid composition and the cryostability of the plasma membrane. J. Plant Res.112: 245–254.
18. Matteucci M, D'Angeli S, Errico S, Lamanna R, Perrotta G, and Altamura MM. 2011. Cold affects the transcription of fatty acid desaturases and oil quality in the fruit of *Olea europaea L.* Genotypes with different cold hardiness. J. Exp. Bot. vol. 62: 3403–3420.

19. Seo PJ, Kim MJ, Park JY, Kim SY, Jeon J, Lee YH, Kim J and Park CM. 2010. Cold activation of a plasma membrane-tethered NAC transcription factor induces a pathogen resistance response in Arabidopsis. Plant J. 61: 661–671.
20. Knight H, Brandt S and Knight MR. 1998. A history of stress alters drought calcium signalling pathways in Arabidopsis. Plant J. 16: 681–687.
21. Ruelland E and Zachowsk A. 2010. How plants sense temperature. Environ. Exp. Bot. 69: 225–232.
22. Zhang S, Jiang H, Peng S, Korpelainen H, and Li C. 2011. Sex-related differences in morphological, physiological, and ultrastructural responses of *Populus cathayana* to chilling. J. Exp. Bot. 62: 675– 686.
23. Sharma Shashi K. and Kumar A. 2015. Role of biomolecules in sensing and signal transducing and acclimation in plants against frost induced low temperature stress – A Review. Int. Res. J. Engg. & Tech. 2(9): 2187-2207.
24. Orvar BL, Sangwan V, Omann F and Dhindsa RS. 2000. Early steps in cold sensing by plant cells: The role of actin cytoskeleton and membrane fluidity. Plant J. 23: 785–794.
25. Saijo Y, Hata S, Kyozuka J. Shimamoto K and Izui K. 2000. Over-expression of a single Ca2+ dependent protein kinase confers both cold and salt/drought tolerance on rice plants. Plant J. 23: 319–327.
26. Townley HE and Knight MR. 2002. Calmodulin as a potential negative regulator of Arabidopsis COR gene expression. Plant Physiol. 128: 1169–1172.
27. Huang C, Ding S, Zhang H, Du H and An L. 2011. CIPK7 is involved in cold response by interacting with CBL1 in *Arabidopsis thaliana*. Plant Sci.181: 57–64.
28. Doherty CJ, Van Buskirk HA and Myers SJ, Thomashow MF.2009. Roles for Arabidopsis CAMTA transcription factors in cold regulated gene expression and freezing tolerance. Plant Cell. 21: 972–984.
29. Shinozaki K and Yamaguchi-Shinozaki K.2000. Molecular responses to dehydration and low temperature: differences and cross-talk between two stress signalling pathways. Curr. Opin. Plant Biol. 3: 217– 223.
30. Shinozaki K, Yamaguchi-Shinozaki K and Seki M. 2003. Regulatory network of gene expression in the drought and cold stress responses. Curr. Opin. Plant Biol. 6: 410–417.
31. Mahajan S and Tuteja N. 2005. Cold, salinity and drought stresses: an overview. Arch. Biochem. Biophys. 444: 139–158.
32. Yamaguchi-Shinozaki K and Shinozaki K. 2006. Transcriptional regulatory networks in cellular responses and tolerance to dehydration and cold stresses. Ann. Rev. Plant Biol. 57: 781–803.
33. Meijer HJ and Munnik T. 2003. Phospholipid-based signalling in plants. Ann. Rev. Plant Biol. 54: 265–306.
34. Munnik T. 2001. Phosphatidic acid: an emerging plant lipid second messenger. Trends Plant Sci. 6: 227–233.
35. Testerink Cand Munnik T. 2005. Phosphatidic acid: a multifunctional stress signalling lipid in plants. Trends Plant Sci. 10: 368–375.
36. Verslues PE and Zhu JK. 2005. Before and beyond ABA: upstream sensing and internal signals that determine ABA accumulation and response under abiotic stress. Biochem. Soc. T. 33: 375–379.
37. Laloi C, Apel K and Danon A. 2004. Reactive oxygen signalling: the latest news,‖ Curr. Opin. Plant Biol. 7: 323–328.
38. Jeon J, Kim NY, Kim, S Kang NY, Novak O, Ku SJ et al. 2010. A subset of cytokinin two-component signalling system plays a role in cold temperature stress response in Arabidopsis. J. Biol. Chem. 285: 23371–23386.

39. Steponkus PL, Uemura M and Webb MS. 1993. A contrast of the cryostability of the plasma membrane of winter rye and spring oat-two species that widely differ in their freezing tolerance and plasma membrane lipid composition. In: Steponkus P.L. (Ed.), Advances in Low-Temperature Biology, Vol. 2, JAI Press, London, pp. 211– 312.
40. Tasseva G, Davy de Virville J, Cantrel C, Moreau F and Zachowski A. 2004. Changes in the endoplasmic reticulum lipid proprieties in response to low temperature in *Brassica napus.* Plant Physiology and Biochemistry. 42: 811-822.
41. De Palma M, Grillo S, Massarelli I, Costa A, Balogh G, Vigh L and Leone A. 2008. Regulation of desaturase gene expression, changes in membrane lipid composition and freezing tolerance in potato plants. Molecular Breeding. 21: 15-26.
42. Badea C and Basu SK. 2009. The effect of low temperature on metabolism of membrane lipids in plants and associated gene expression. Plant Omics Journal. 2: 78-84.
43. Shanklin J and Cahoon ES.1998. Desaturation and related modifications of fatty acids. Ann. Rev. Plant Physiol. Plant Mol. Biol. 49: 611- 641.
44. Wada H and Murata N. 1998. Membrane lipids in cyanobacteria In Lipids in Photosynthesis: Structure, Function and Genetics, Edited by Siegenthaler PA, Murata N. Dordrecht: Kluwer Academic Publishers, pp. 65-81 .
45. Los DA and Murata N. 1998. Structure and expression of fatty acid desaturases. Biochim Biophys Acta. 1394: 3-15.
46. Suzuki N, Koussevitzky S, Mittler R and Miller G. 2011. ROS and redox signalling in the response of plants to abiotic stress. Plant Cell Environ. 35: 259–270.
47. Gechev T, Willekens H, Van Montagu M, Inze D, Van Camp W, Toneva V and Minkov I. 2003. Different responses of tobacco antioxidant enzymes to light and chilling stress. J. Plant Physiol. 160: 509–515.
48. Lane N. 2002. Oxygen: The Molecule that Made the World. Oxford University Press.
49. Bielski BHJ, Arudi RL, and Sutherland MW. 1983. A study of the reactivity of HO2/ O2- with unsaturated fatty acids. Journal of Biological Chemistry. 258: 4759–4761.
50. Browne RW and Armstrong D. 2000. HPLC analysis of lipidderived polyunsaturated fatty acid peroxidation products in oxidatively modified human plasma. Clinical Chemistry. 46: 829–836.
51. Schneider C, Boeglin WE, Yin H, Porter NA, and Brash AR, 2008. Intermolecular peroxyl radical reactions during autoxidation of hydroxy and hydroperoxy arachidonic acids generate a novel series of epoxidized products. Chemical Research in Toxicology. 21: 895–903.
52. Yin H, Xu L, and Porter NA. 2011. Free radical lipid peroxidation: mechanisms and analysis. Chemical Reviews. 111: 5944–5972.
53. Shinozaki K and Yamaguchi-Shinozaki K.2000. Molecular responses to dehydration and low temperature: differences and cross-talk between two stress signalling pathways. Curr. Opin. Plant Biol. 3: 217– 223.
54. Liu Q, Kasuga M, Sakuma Y, Abe H, Miura S, Yamaguchi-Shinozaki K and Shinozaki K. 1998. Two transcription factors, DREBI and DREB2, with an EREBP/AP2 DNA binding domain separate two cellular signal transduction pathways in drought and low temperature responsive gene expression respectively, in Arabidopsis. Plant Cell. 10: 1391–1406.
55. Frankow-Lindberg B. E. 2001. Adaptation to winter stress in nine white clover populations: changes in non-structural carbohydrates during exposure to simulated winter conditions and spring regrowth potential. Ann. Bot. 88: 745–751.

56. Hernández-Nistal J, Dopico B and Labrador E. 2002. Cold and salt stress regulates the expression and activity of a chickpea cytosolic Cu/Zn superoxide dismutase. Plant Sci. 163: 507–514.
57. Baek KH and Skinner DJ. 2003. Alteration of antioxidant enzyme gene expression during cold acclimation of near-isogenic wheat lines. Plant Sci. 165: 1221–1227.
58. Goulas E, Schubert M., Kieselbach T, Kleczkowski LA, Gardestrom P, Schroder W and Hurry V. 2006. The chloroplast lumen and stromal proteomes of Arabidopsis thaliana show differential sensitivity to short- and long-term exposure to low temperature. Plant J. 47: 720–734.
59. Baena-Gonzalez E, Gray JC, Tyystjarvi E, Aro EM and Maenpaa P. 2001. Abnormal regulation of photosynthetic electron transport in a chloroplast ycf9 inactivation mutant. J. Biol. Chem. 276: 20795–20802.
60. Krol M, Ivanov AG, Jansson S, Kloppstech K and. Huner NP. 1999. Greening under high light or cold temperature affects the level of xanthophyll-cycle pigments, early light-inducible proteins, and light-harvesting polypeptides in wild-type barley and the Chlorina f2 mutant. Plant Physiol. 120: 193–204.
61. Ruelland E , Vaultier MN, Zachowski A, Hurry V, Kader JC, and Delseny M. 2009. Cold signalling and cold acclimation in plants. Adv. Bot. Res. 49: 35–150.
62. Ivanov AG, Sane PV, Krol M, Gray GR, Balseris A, Savitch LV, Oquist G and Huner NP. 2006. Acclimation to temperature and irradiance modulates PSII charge recombination. 2006. FEBS Lett. 580: 2797–2802.
63. Passarini F, Wientjes E, Hienerwadel R and Croce R. 2009. Molecular basis of light harvesting and photoprotection in CP24: unique features of the most recent antenna complex. J. Biol. Chem. 284: 29536–29546.
64. Han H, Gao S, Li B, Dong XC, Feng HL, Meng QW. 2010. Overexpression of violaxanthin de-epoxidase gene alleviates photoinhibition of PSII and PSI in tomato during high light and chilling stress. J. Plant Physiol . 167: 176–183.
65. Hekneby M, Antolı´n MC and Sa´nchez-Dı´az M. 2006.Frost resistance and biochemical changes during cold acclimation in different annual legumes. Environ. Exp. Bot. 55: 305–314.
66. Patton AJ, Cunningham SM, Volenec JJ and Reicher ZT. 2007. Differences in freeze tolerance of zoysiagrasses: II. Carbohydrates and proline accumulation. Crop Science Society of America, Madison.
67. Fernandez O, Theocharis A, Bordiec S, Feil R, Jacquens L, Cle´ment C, Fontaine F, Ait Barka E. 2012. *Burkholderia phytofirmans* strain PsJN acclimates grapevine to cold by modulating carbohydrates metabolism. Mol. Plant Microbe Interact. 25: 496–504.
68. Welling A and Palva ET. 2006. Molecular control of cold acclimation in trees. Physiol Plant. 127:167–181.
69. Uemura M, Warren G, Steponkus PL. 2003. Freezing sensitivity in the sfr4 mutant of Arabidopsis is due to low sugar content and is manifested by loss of osmotic responsiveness. Plant Physiol. 131:1800–1807.
70. Bolouri-Moghaddam MR, Le Roy K, Xiang L, Rolland F, and Van den Ende W. 2010. Sugar signalling and antioxidant network connections in plant cells. FEBS J. 277: 2022–2037.
71. Keunen E., Peshev D, Vangronsveld J, Van den Ende W and Cuypers A. 2013. Plant sugars are crucial players in the oxidative challenge during abiotic stress: extending the traditional concept. Plant Cell Environ.36: 1242–1255.
72. Valluru R, Lammens W, Claupein W and Van den EndeW. 2008. Freezing tolerance by vesicle-mediated fructan transport. Trends Plant Sci.13: 409–414.

73. Stitt M and Hurry V. 2002. A plant for all seasons: alterations in photosynthetic carbon metabolism during cold acclimation in Arabidopsis. Curr. Opin. Plant Biol. 5: 199–206.
74. Bohnert HJand Sheveleva E. 1998. Plant stress adaptations — making metabolism move. Curr. Opin. Plant Biol. 1: 267–274.
75. Williams LE, Lemoine R and Sauer N. 2000. Sugar transporters in higher plants—a diversity of roles and complex regulation. Trends Plant Sci. 5: 283–290.
76. Couee I, Sulmon C, Gouesbet G, and El Amrani A. 2006. Involvement of soluble sugars in reactive oxygen species balance and responses to oxidative stress in plants. J. Exp. Bot.57: 449–459.
77. Aı¨t Barka E and Audran JC. 1996. Response des vignes champenoises aux tempe´ratures ne´gatives: effect d'un refroidissement contro^le´ sur les re´serves glucidiques du complexe gemmaire avant et au cours du de´bourrement. Can. J. Bot. 74: 492–505.
78. Tabaei-Aghdaei SR, Pearce RS and Harrison P. 2003. Sugars regulate cold-induced gene expression and freezing-tolerance in barley cell cultures. J. Exp. Bot. 54: 1565–1575.
79. Fernandez O, Bethencourt L, Quero A, Sangwan RS, and Clement C. 2010. Trehalose and plant stress responses: friend or foe,Trends Plant Sci.15: 409–417.
80. Pollock CJ and Lloyd EJ. 1987. The effect of low temperature upon starch, sucrose and fructan synthesis in leaves. Ann. Bot. 60: 231–235.
81. Bohnert HJ and Sheveleva E. 1998. Plant stress adaptations—making metabolism move. Curr. Opin. Plant Biol. 1: 267–274.
82. Verbruggen N and Hermans C. 2008. Proline accumulation in plants: a review. Amino Acids. 35: 753–759.
83. Szabados L and Savoure A. 2010. Proline: a multifunctional amino acid. Trends Plant Sci. 15: 89–97.
84. Kaur G, Kumar S, Thakur P, Malik JA, Bhandhari K, Sharma KD and Nayyar H. 2011. Involvement of proline in response of chickpea (*Cicer arietinum L.*) to chilling stress at reproductive stage. Sci. Hortic. 128: 174–181.
85. Chen TH and Murata N. 2008. Glycinebetaine: an effective protectant against abiotic stress in plants. Trends Plant Sci. 13: 499–505.
86. Sakamoto A and Murata N. 2002. The role of glycine betaine in the protection of plants from Stress: clues from transgenic plants. Plant Cell Environ. 25: 163–172.
87. Rhodes D and Hanson AD. 1993.Quaternary ammonium and tertiary sulfonium compounds in higher plants. Ann. Rev. Plant Physiol. & Plant Mol. Biol. 44: 357–384.
88. Lee CB, Hayashi H, Moon BY. 1997. Stabilization by glycinebetaine of photosynthetic oxygen evolution by thylakoid membranes from Synechococcus PCC7002. Mol. Cells. 7: 296–299.
89. McNeil SD, Nuccio ML, HansonAD. 1999. Betaines and related osmoprotectants,Targets for metabolic engineering of stress resistance. Plant Physiol. 120: 945–949.
90. Chen WP, Li PH and Chen THH. 2000. Glycinebetaine increases chilling tolerance and reduces chilling-induced lipid peroxidation in *Zea mays L.* Plant Cell Environ. 23: 609–618.
91. Alia Hayashi H, Chen THH and Murata N. 1998. Transformation with a gene for choline oxidase enhances the cold tolerance of Arabidopsis during germination and early growth. Plant Cell Environ. 21: 232– 239.
92. Hématy K, Cherk C and Somerville S. 2009. Host-pathogen warfare at the plant cell wall. Curr. Opin. Plant Biol.12: 406–413.
93. Szymanski DB and Cosgrove DJ. 2009. Dynamic coordination of cytoskeletal and cell wall systems during plant cell morphogenesis. Curr. Biol.19: 800–811.

94. Denness L, McKenna JF, Segonzac C, Wormit A and Madhou P et al. 2011. Cell wall damage-induced lignin biosynthesis is regulated by a Reactive Oxygen Species-and Jasmonic Acid- dependent process in Arabidopsis. Plant Physiol. 156:1364–1374.
95. Temple BRS and Jones AM. 2007. The plant heterotrimeric G-protein complex. Ann. Rev. Plant Biol. 58: 249–266.
96. Seifert GJ and Blaukopf C. 2010. Irritable walls: the plant extracellular matrix and signalling. Plant Physiol. 153: 467–478.
97. Tsang DL, Edmond C, Harrington JL and Nühse TS. 2011. Cell wall integrity controls root elongation via a general 1-Aminocyclopropane-1-Carboxylic Acid-dependent, Ethylene-independent pathway. Plant Physiol. 156: 596–604.
98. Taylor-Teeples M, Lin L, de Lucas M, Turco G, Toal TW et al. 2015. An Arabidopsis gene regulatory network for secondary cell wall synthesis. Nature. 517: 571–575.
99. Ellis C, Karafyllidis, Wasternack Cand Turner JG. 2002. The Arabidopsis mutant cev1 links cell wall signalling to jasmonate and ethylene responses. Plant Cell.14: 1557–1566.
100. Manfield IW, Orfila C, Mccartney L, Harholt J, Bernal AJ et al. 2004. Novel cell wall architecture of isoxaben- habituated Arabidopsis suspension-cultured cells: global transcript profiling and cellular analysis. Plant J. 40: 260–275.
101. Hamann T, Bennett M, Mansfield J and Somerville C. 2009. Identification of cell-wall stress as a hexose dependent and osmosensitive regulator of plant responses. Plant J. 57: 1015–1026.
102. Le MQ, Pagter M and Hincha DK. 2015. Global changes in gene expression, assayed by microarray hybridization and quantitative RT-PCR, during acclimation of three Arabidopsis thaliana accessions to sub-zero temperatures after cold acclimation. Plant Mol. Biol. 87: 1–15.
103. Yamada T, Kuroda K, Jitsuyama Y, Takezawa D, Arakawa K and Fujikawa S. 2002. Roles of the plasma membrane and the cell wall in the responses of plant cells to freezing. Planta. 215: 770–778.
104. Ferrer JL, Austin MB, Stewart Jr C and Noel JP. 2008. Structure and function of enzymes involved in the biosynthesis of phenylpropanoids. Plant Physiol. Biochem. 46: 356–370.
105. Shafi A, Dogra V, Gill T, Ahuja PS and Sreenivasulu Y. 2014. Simultaneous over-expression of PaSOD and RaAPX in transgenic Arabidopsis thaliana confers cold stress tolerance through increase in vascular lignifications. PLoS One. 9: e110302.
106. Raes J, Rohde A, Christensen JH, Van de Peer Y and Boerjan W. 2003. Genome-wide characterization of the lignification toolbox in Arabidopsis. Plant Physiol. 133: 1051–1071.
107. Olsen KM, Lea US, Slimestad R, Verheul M and Lillo C. 2008. Differential expression of four Arabidopsis PAL genes; PAL1 and PAL2 have functional specialization in abiotic environmental-triggered flavonoid synthesis.J. Plant Physiol.165: 1491–1499.
108. Cass CL, Peraldi A, Dowd PF, Mottiar Y, Santoro N et al. 2015. Effects of phenylalanine ammonia lyase (PAL) knockdown on cell wall composition, biomass digestibility, and biotic and abiotic stress responses in *Brachypodium.* J. Exp. Bot. 66: 4317–4335.
109. Ezaki B, Sasaki K, Matsumoto H and Nakashima S. 2005. Functions of two genes in aluminium (Al) stress resistance: repression of oxidative damage by the AtBCB gene and promotion of efflux of Al ions by the NtGDI1gene. J. Exp. Bot. 56: 2661–2671.
110. Hongtao J, Wang Y, Cloix C, Li K, Jenkins GI, Wang S et al. 2015. The Arabidopsis RCC1 Family Protein TCF1 Regulates Freezing Tolerance and Cold Acclimation through Modulating Lignin Biosynthesis. PLoS Genet. 11: e1005471.

111. Fitter AH, Hay RKM (2002) Environmental physiology of plants, 3rd edn. Academic, London.
112. Zróbek-sokolnik, A. 2012. Temperature stress and responses of plants. In: Ahmad P, Prasad MNV (eds) Environmental adaptations and stress tolerance of plants in the era of climate change. New York: Springer. Pp. 113-134.
113. Hasanuzzaman M, Nahar K and Fujita M. 2013. Extreme Temperature Responses, Oxidative Stress and Antioxidant Defense in Plants. In: Abiotic Stress – Plant Responses and Applications in Agriculture. https://www.intechopen.com/books/abiotic-stress-plant-responses-and-applications-in-agriculture/extreme-temperature-responses-oxidative-stress-and-antioxidant-defense-in-plants
114. Hu WH, Song XS, Shi K, Xia XJ, Zhou YH and Yu, J Q. 2008. Changes in electron transport, superoxide ismutase and ascorbate peroxidase isoenzymes in chloroplasts and mitochondria of cucumber leaves as influenced by chilling. Photosynthetica. 46: 581-588.
115. Kotak S, Larkindale J, Lee U, Von Koskull-döring P, Vierling E, and Scharf KD. 2007. Complexity of the heat stress response in plants. Current Opinion in Plant Biology.10:310-316.
116. Davanport TL. 2009. Reproductive physiology. In: Litz RE (ed) The Mango, 2nd edn. Botany, Production and Uses. CABI: London.
117. Tindall HD. 1994. Rambutan Cultivation. Food and Agriculture Organization of the United Nations. FAO: Rome. 978-9-25103-325-8.
118. Verheij EWM and Coronel RE. 1992. Edible fruits and nuts. Plant Resources of South-East Asia Prosea Foundation: Bogor. 2: 128-131.
119. Viktorova MK. 1983. Plant growing in the tropics and subtropics. Mir publisher:, Moscow.214-218.
120. Larcher W. 1995. Physiological Plant Ecology-Ecophysiology and Stress Physiology of Functional Groups. Berlin: Springer.
121. Haq N. 2006. Fruits for the Future 10. Jackfruit (*Artocarpus heterophyllus*). Southampton Centre for Underutilised Crops. 192p.
122. Whiley AW, Rasmussen TS, Saranah JB, and Wolstenholme BN. 1989. Effect of temperature on growth, dry matter production and starch accumulation in ten mango (*Mangifera indica* L.) cultivars. Journal of Horticultural Science and Biotechnology. 64: 753-765.
123. Dinesh MR and Reddy BMC. 2012. Physiological basis of growth and fruit yield characteristics of tropical and Sub-tropical fruits to temperature. *In*: Tropical fruit tree species and climate change. Eds.: Sthapit BR, Ramanatha Rao V, Sthapit S. Biodiversity International, New Delhi.
124. Sharma Shashi K. 2016. Freeze Restriction Studies for Frost Protection in Subtropical Fruit Plants. Project Report (HM-218-62). Department of Science and Technology (DST/SSTP/HP/170/2011.
125. Soliemani A, Lessani H and Talaie A. 2003. Relationship between stomatal density and ionic leakage as indicators of cold hardiness in olive (*Olea europea* L.). *Acta Hort.* 618: 521-525.
126. Sharma Shashi K. 2015. Effect of surface wetness and duration of low temperature exposure on frost damage in subtropical fruit species. *The Asian J. Hort.* 10: 272-277.
127. Sharma Shashi K. 2012. Studies on visualizing frost/freeze damage in subtropical fruit species. *Indian J. Hort.* 69(1): 27-32.
128. Levitt J. 1980. Responses of plants to environmental stresses, Vol. 1. New York:

Academic Press.

129. Wang CY and Wallace HA. 2003. Chilling and freezing injury. *In:* K.C Gross, C.Y. Wang and M. Saltveit (eds). *The Commercial Storage of Fruits, Vegetables, and Florist and Nursery Stocks.* USDA Handbook Number, No. 66. See: http://www.ba.ars.usda.gov/hb66/index.html
130. Haldimann P. 1998. Low growth temperature-induced changes to pigment composition and photosynthesis in *Zea mays* genotypes differing in chilling sensitivity. Plant Cell and Eviron. 21: 200–208. https://onlinelibrary.wiley.com/ doi/epdf/10.1046/j.1365-3040.1998.00260.x
131. Dexter, S. T., Tottingham, W. E. and Garber, L. F. 1932. Investigations of the hardiness of plants by measurement of electrical conductivity. Plant Physiology. 7: 63-78. doi: 10.1104/pp.7.1.63
132. Bartolozzi, F. and Fontanazza, G. 1999. Assessment of frost tolerance in olive (*Olea europaeae* L.). *Scientia Hort.***81**: 309-319.
133. Mancuso S 2002 Electrical resistance changes during exposure to low temperature measure chilling and freezing tolerance in olive tree (*Olea europea* L.) plants. *Plant, Cell and Environment*. 23:291-299.
134. Barker R, Fricker R, Dunnet, SB. 1994. 14 - Factors Important in the Survival of Dopamine Neurons in Intracerebral Grafts of Embryonic Substantia Nigra. In: Methods in Neurosciences. Vol. 21. 237-252pp. Elsevier.
135. Fiorino, P. and Mancuso S. 2000. Olivo e basse temperature: Danni da freddo, adattamento e resistenza L' Informatore Agraio 22: 55-59.
136. Pearce, R. S. and Fuller, M. P. 2001. Freezing of barley studied by infrared video thermography. Plant Physiology. 125: 227-240.
137. Endoh K, Kasuga J, Arakawa K, Fujikawa S 2009. Cryo-scanning electron microscopic study on freezing behaviors of tissue cells in dormant buds of larch (Larix kaempferi). Cryobiology 59:214–222.
138. Ishikawa, M., 1984 - Deep super-cooling in most tissues of wintering *Sasa senanensis* and its mechanism in leaf blade tissues. Plant Physiol.75: 196-202.
139. Bartolozzi F and Fontanazza G. 1999. Assessment of frost tolerance in olive (Olea europea L.). Scientia Hort. 81: 309-319.
140. Fiorino, P. and Mancuso, S. 2000. Differential thermal analysis, super-cooling and cell viability in organs of *Olea europaea* at subzero temperatures. Advances in Horticultural Science. 14(1): 23-27.
141. Ishikawa, M. Price, W. S., Ide, H. and Arata, Y. 1997. Visualization of freezing behaviours in leaf and flower budsof full monn maple nuclear magnetic resonance microscopy. Plant Physiol. **115**: 1515-1524.
142. Price, W. S., Ide, H. Arata, Y. and Ishikawa, M. 1997. Visualization of freezing behavior in leaf bud tissues of *Rhododendron japonicum* by nuclear magnetic resonance micro imaging. Australian J. Plant Physiol. 24: 599-605.
143. Ide, H., Price, W. S., Arata, Y. and Ishikawa, M. 1998. Freezing behaviours in leaf buds of cold hardy conifers vilualized by NMR microscopy. Tree Physiol. 18: 451-458.
144. Sinclair BJ, Gibbs AG, Lee W-K, Rajamohan A, Roberts SP, Socha JJ. 2009. Synchrotron X-Ray Visualisation of Ice Formation in Insects during Lethal and on-Lethal Freezing. PLoS ONE 4(12): e8259. https://doi.org/10.1371/journal.pone.0008259
145. Kovaleski AP, Londo JP and Finkelstein KD. 2019. X-ray phase contrast imaging of Vitis spp. buds shows freezing pattern and correlation between volume and cold hardiness. Scientific Reports.9: 14949. https://doi.org/10.1038/s41598-019-51415-2

146. Wisniewski, M., Lindow, S. E. and Ashworth, E. N. 1997. Observations of ice nucleation and propagation in plants using infrared video thermography. Plant Physiol. 113: 327-324
147. Pearce, R. S. 1988. Plant freezing and damage. Annals of Botany. 87: 417-424.
148. Rozante JR, Gutierrez ER, Dias PLDS, Farnanades A de A, Alvim DS, Silva VM. 2019. Development of an index for frost prediction: Technique and validation. Meteorological Applications. 27(1):1-12. DOI: https://doi.org/10.1002/met.1807
149. Krasovitski, B., Kimmel, E., Rozenfeld, M., Amir, I. 1999. Aqueous foams for frost protection of plants: stability and protective properties. *J. Agric. Engng. Res.* 72:177-185.
150. Sharma, Shashi K. 1997. Yield forecasting in apple based on meteorological, morphological, anatomical and nutritional parameters. Ph.D. Thesis. College of Horticulture, DR YSP UHF, Solan (HP) 173230 India.
151. Theil, H. 1966. Applied Economic Forecasting. Pp. 26-36, North Holland.
152. Anonymous. 2020. Patterns of wind, water and weather in the tropics. The Pennsylvania State University. https://www.e-education.psu.edu/meteo3/l11_p2.html
153. Kalma, J. D., Laughlin, G. P., Caprio, J. M. and Hamer, P. J. C. 1992. Remote Sensing for Frost Risk Mapping and Frost Prediction. In: The Bioclimatology of Frost. pp 61-65.
154. Sharma, Shashi K. and Badiyala, S. D. 2008. Prioritization of sub-tropical fruit plants for frost prone low hill region of Himachal Pradesh. *Natural product Radiance*. 7:347-353.
155. Sharma, Shashi K., Verma, K. S., Preet Pratima and Banyal A. 2014. Delineation of Frost Prone Agro-ecological Situations and Fruit Crops Prioritization for the Subtropical Himalayan Region of Himachal Pradesh. International Journal of Latest Technology in Engineering, Management and Applied Science. III(X): 69-74.
156. Attaway, J.A. 1997. A history of Florida citrus freezes. Lake Alfred, Florida: Florida Science Source, Inc.
157. Bagdonas, A., Georg, J.C. & Gerber, J.F. 1978. Techniques of frost prediction and methods of frost and cold protection. World Meteorological Organization Technical Note, No. 157. Geneva, Switzerland. 160p.
158. Braud, H. J. and Jerry, L.C. 1970. Physical properties of foam for protecting plants against cold weather. *Trasactions of Americal Soc. Of Agric. Engineers*. 13:1-5.
159. Siminovich, D., Rheaume, B., Lyall, I. H. and Buttler, J. (1972). Foam for frost protection of crops. Canada Department of Agriculture Publication 1490. Agriculture Canada.
160. Anonymous. 2018. Oranges that grow in cold climate. http://homeguides.sfgate.com/orange-trees-grow-cold-climate-55135.html
161. Burns, R. 2014. Super cold hardy Satsuma orange frost named Texas Superstar. Agrilife Today. http://agrilifetoday.tamu.edu/2014/06/04/satsuma-orange-frost-texas-superstar/
162. Grant A. 2018. Zone 7 Citrus trees: Tips on growing citrus trees in Zone 7. http://www.gardeningknowhow.com/garden-how-to/gardening-by-zone/zone-7/zone-7-citrus-trees.htm .
163. Bruke, M. J. Gust, L. V., Quamme, H. A., Weiser, C. J., Li, P.H.1976. Freezing and Injury in Plants. Ann. Rev. Physiol. 27: 507-528.
164. Li, P. H. and Plata, J.P. 1978. Frost hardening and freezing stress in tuber bearing *Solanum* species. In: Plant Cold Hardiness and Freezing Stress: Mecanism and Crop Implications. Li, P. H. , Sakai, A. eds. Academic Press New York. pp. 49-71.
165. Gusta, L.V., Bruke, M. J. and Kapoor, A. C. 1975. Determination of unfrozen water in winter cereals at subfreezing temperatures. Plant Physiology. 56(5): 707-709.

166. Arias, N. S., Bucci, S. J., Scholz, F. G. and Goldstein, G. 2015. Freezing avoidance by super-cooling in *Olea europea* cultivars: the role of apoplastic water. Solute content and cell wall rigidity. Plant, Cell & Environment. 38(10). DOI: http://doi.org/10.1111/pce.12529
167. Gusta, L. V. and Wisniewski, M. 2012. Understanding plant cold hardiness: An opinion. *Physiologia Planatarum*. 147(1): DOI: http://doi.org/10.1111/j.1399-3054.2012.01611.x
168. Quellet F. CharronJ-B.2013. Cold acclimation and freezing tollerence in plants. Ib:eLS. John Wiley & Sons, Ltd.: Chichester. DOI: 10.1002/9780470015902.a0020093.pub2
169. Sharma Shashi K. and Preet Pratima. 2018. Seasonal variation in leaf water, lipid content and the corresponding cold acclimation in subtropical fruit plants. *International Journal of Current Research*. 10(3): 66723-66727.
170. Steponkus PL (1984) Role of the plasma membrane in freezing injury and cold acclimation. Ann. Rev. Plant Physiol. 35: 543–584.
171. Cooper GM. 2000. Structure of Plasma Membrane. In: The Cell: A Molecular Approach. 2nd Ed. Sunderland (MA): Sinauer Associates.
172. Sanders D, Pelloux J, Brownlee C and Harper JF. 2002. Calcium at the crossroads of signaling. Plant Cell. 14(Suppl): S401–S417.
173. Du L, Poovaiah BW. 2005. Ca2+/calmodulin is critical for brassinosteroid biosynthesis and plant growth. Nature. 437: 741–745.
174. Sangwan V, Foulds I, Singh J and Dhindsa RS 2001. Cold-activation of Brassica napus BN115 promoter is mediated by structural changes in membranes and cytoskeleton, and requires Ca2+Influx. Plant J. 27: 1–12.
175. Catala R, Santos E, Alonso JM,. Ecker JR, Martinez-Zapater JM and Salinas J. 2003. Mutations in the Ca2?/H? transporter CAX1 increase CBF/DREB1 expression and the cold-acclimation response in Arabidopsis. Plant Cell. 15: 2940–2951.
176. Lane N. 2002. Oxygen: The Molecule that Made the World. Oxford University Press, 2002
177. Halliwell B. and Gutteridge J.M.C. 1984. Oxygen toxicity, oxygen radicals, transition metals and disease. Biochemical Journal. 219. 1: 1–14.
178. Bielski BHJ, Arudi RL and. Sutherland MW. 1983. A study of the reactivity of HO_2/O_2- with unsaturated fatty acids, Journal of Biological Chemistry. 258 (8): 4759–4761.
179. Browne R W and Armstrong D. 2000. HPLC analysis of lipid derived polyunsaturated fatty acid peroxidation products in oxidatively modified human plasma. Clinical Chemistry. 46 (6- 1): 829–836.
180. Powell AA and Himelrick DG. 2000. Principles of freeze protection for fruit crops. Alabama Cooperative Extension System, ANR 1057B. http://www.aces.edu.
181. Mee TR and Bartholic JF. 1979. Man-made fog. pp. 334-352, in: Barfield and Gerber, 1979, q.v.
182. Oude Vrielink AS, Aloi A, Olijve LLC and Voets IK. 2016. Interaction of ice binding proteins with ice, water and ions. Biointerphases, 11(1), [018906]. https://doi.org/10.1116/1.4939462
183. Groenzin H, Li I and Shultz MJ. 2008. Sum-frequency generation: polarization surface spectroscopy analysis of the vibrational surface modes on the basal face of ice Ih. J. Chem. Phys. 128. 214510, 1- 214510, 8.
184. Wilson PW. 1994. A model for thermal hysteresis utilizing the anisotropic interfacial energy of ice crystals. Cryobiology. 31: 406–412.
185. Jorov A, Zhorov BS and Yang DS. 2004. Theoretical study of interaction of winter flounder antifreeze protein with ice. Protein Sci. 13:1524–1537.

186. Madura JD, Dalal P and Haymet T. 2003. Interfacial simulations. Symposium on stress proteins: From antifreeze to heat shock. March 7–9, 2003. University of California Davis, Bodega Bay Bay, California.
187. Buckley SL and Lilliford PJ. 2009. Antifreeze proteins: their structure, binding and use. In Modern Biopolymer Science (S. Kasapis, I.T. Norton and J.B. Ubbink, eds.) pp. 93–128, Elsevier, London.
188. Glenn DM, Puterka G, Baugher T, Unruh T and Drake S. 1998. Hydrophobic particle films improve tree fruit productivity. HortScience. 33: 547.
189. Sapienza R. 2001. The carbohydrates and the amino acid salts formed also are useful components of a deicing agent. U.S. Pat. No. 5,876,621.
190. Fuller MP, Hamed F, Wisniewski M, and Glenn DM . 2003. Protection of plants from frost using hydrophobic particle film and acrylic polymer. Annals of Applied Biology. 143: 93-98. doi.org/10.1111/j.1744-7348.2003.tb00273.x
191. Wisniewski M, Fuller M, Glenn D M, Palta J, Carter J, Gustta L, Griffith M and Duman J 2001 Factors involved in ice nucleation and propagation in plants: an overview based on new insights gained from the use of infrared thermography. *Buvisindi –Icel. Agr. Sci.* 14: 41-47.
192. Anonymous. 2021. Correct and simple whitewashing of trees and bushes. Easy way to have more healthier crop. https://botanchik.ru/en/gardeners-page/382-correct-and-simple-whitewashing
193. Anonymous. 2018. Cleaning and Disinfecting Pruning Tools for Orchard Crops. Integrated Pest management. https://ipm.missouri.edu/MEG/2018/1/cleaning_pruning_tools/
194. Sharma Shashi K. 2014. Growth restoration studies in frost-affected mango (*Mangifera indica* L.) orchards in the sub-Himalayan region. *Journal of Horticultural Science*. 9(1): 12-17.